D0758649

Reasoning in Physics

Reasoning in Physics

The Part of Common Sense

by

Laurence Viennot

Université Denis Diderot (Paris 7), France

KLUWER ACADEMIC PUBLISHERS

DORDRECHT / BOSTON / LONDON

A C.I.P. Catalogue record for this book is available from the Library of Congress.

ISBN 0-7923-7140-2

Published by Kluwer Academic Publishers,
P.O. Box 17, 3300 AA Dordrecht, The Netherlands.

Sold and distributed in North, Central and South America
by Kluwer Academic Publishers,
101 Philip Drive, Norwell, MA 02061, U.S.A.

In all other countries, sold and distributed
by Kluwer Academic Publishers,
P.O. Box 322, 3300 AH Dordrecht, The Netherlands.

Printed on acid-free paper

Original publication:
Raisonner en physique. La part du sens commun.
Published by: De Boeck & Larcier s.a., 1996.
Translated by: Amélie Moisy

Printed in the Netherlands.

Table of Contents

Acknowledgments

This work is the result of a long-term group effort, to which also contributed, in our many discussions, Ahmed Fawaz and more recently, Martine Méheut and Gérard Rebmann. My sincere thanks to them. The support and interest of my Physics colleagues at the Université Denis Diderot have also proved essential.

I am indebted to the pupils, students, teachers and university professors without whom the investigations on which this book is based would not have been possible.

I am grateful to Michel Viennot whose careful and demanding study of this book provided me with the views of a non-specialist with a good knowledge of science.

The translation of this work was carried out under excellent conditions thanks to the competence and kindness of Amélie Moisy. And Robin Millar, who read the English version over completely, has been of invaluable aid. My warmest appreciation to them both.

About the Author

Laurence Viennot is a Professor at Denis Diderot University (Paris 7). She teaches Physics and Didactics of Physics. She heads a post-graduate studies programme (DEA) in Didactics and various teacher-training units. She has been a member of the national committee in charge of preparing new curricula in Physics (GTD) for secondary schools in France (1990-1995), and a member of the first executive board of the European Science Education Research Association (ESERA), founded in 1995. This book is mainly based on studies conducted by the author's research team (Laboratoire de Didactique de la Physique dans l'Enseignement Supérieur, now Laboratoire de Didactique des Sciences Physiques). Abdelmadjid Benseghir Helena Caldas, Françoise Chauvet, Jean-Louis Closset, Wanda Kaminski,, Laurence Maurines, Jacqueline Menigaux, Sylvie Rainson, Sylvie Rozier and Edith Saltiel have contributed to this work.

Preface

Common sense is said to be the best distributed commodity in the world. That should be reassuring, when so many other assets are so unequally shared here on Earth! But this common resource sometimes has dangerous effects. According to the dictionary, however, it is a "form of judgment and action common to all men," a "capacity for correct and dispassionate judgment when problems cannot be resolved by scientific reasoning" (Robert dictionary). That brings us to the very heart of the matter that Laurence Viennot deals with in this book: Do science – in this case, physics – and common sense really occupy two separate areas of thought, as the dictionary so authoritatively states? Obviously not, for scientific knowledge and reasoning – the *scientific mind* described by Bachelard – are the result of a long process of mental organisation, in which the meaning of words gradually changes, clear or shaky concepts are constructed, and one's representations of the world start to differ from those one may have had since birth, or has learnt at school, or simply picked up along the way – in short, from all the things that constitute common sense.

That kind of sense, still common to children today, supposedly leads to good judgment; it cannot accept that the Earth moves around the sun, that people can stand upright in the antipodes, or that light travels from an object to one's eye and not the other way around. Regarding the motion of the Earth, it was only after a lengthy process, from Aristotle to Einstein, that scientific reasoning managed to overcome a misguided intuition based on sense perception, and came to accept relative motion, that of planes, trains or Foucault's pendulum.

The inherent animism of common sense bestows upon concepts (an optical image, the speed of a ball, the magnetic field) characteristics of material objects, and we expect them to behave like a fob-watch, a fork or a grain of sand. What is more, we endow these objects with properties, tendencies, virtues and desires, we imagine that they can feel love or hostility: a patently anthropomorphic view of things, comparable to Aristotle's; twenty-five centuries on, these beliefs still endure, and one must take them seriously. To top things off, these concept-object characters are seen as acting out stories, references to which can be found in pedagogical exchanges between teachers and students, which Laurence Viennot has carefully studied.

When the author writes that "the goal of science is to establish a serious competitor to natural thought, whose coherence and predictive power are clearly superior," she is saying that there must be a major effort towards intellectual lucidity, and setting the goals of physics teaching far beyond the training of future engineers or scientists. For common sense is not only misleading when applied to expanding gases, overlapping rays of light, or bouncing springs – who among us has no doubt at all about how a car runs, how Social Security is financed, how retirement pensions work, or what the greenhouse effect is? Physics tackles an extraordinarily complex reality and proposes pertinent simplifications. Physics *simplifies,* as it extracts from this complexity certain factors that it considers as decisive and measurable: isn't it remarkable how a system (a few cubic decimetres of gas) composed of 10^{23} atoms – in itself a huge number of independent particles – can be described so rigorously by just two quantities, temperature and volume? It is *pertinent*, in that it gives us a means of acting upon the world, and of predicting events, whose limits and strengths are known to us. Is it not possible that we will make more responsible citizens, more enlightened parents, less dysfunctional professionals, and healthier old people, if we learn to go beyond the seemingly evident conclusions of common sense, confront resistant reality as physics teaches us to, and apply this new talent to the innumerable day-to-day occurrences of "civilised" life in which we have to confront extremely complex situations?

Laurence Viennot is a physicist by training, but she is also an academic who has chosen a relatively unpopular field, didactics. "Why are students so bad, when they have such good teachers (us)?": who has never heard this sad refrain, in faculty rooms from kindergarten to university? But, rather than complain, Laurence Viennot has honed her tests and questionnaires, double-checked her hypotheses, made patient investigations at every level of education, compared her findings worldwide – in short, hers has been a life devoted to well-conducted research. The conclusions presented here are not based only on her own studies, but also on those of doctoral students whom she or close colleagues have counselled, and on those of other researchers (particularly at the Laboratoire de Didactique de la Physique dans l'Enseignement supérieur at Paris VII-Denis Diderot University), who, like her, are striving to understand what is going wrong.

Her findings, presented in a lively, humorous, and modest fashion, will no doubt inspire many teachers to re-orientate their pedagogical approaches, and to overcome obstacles to students' understanding which were hitherto thought to be insuperable. Their own view of physics may change – as when Einstein began taking certain questions literally, and imagined following a wave of light by moving at its own speed, or falling with an elevator. Having rid himself of reifying common sense, he was less open to surprise, and new, fertile vistas suddenly opened up before him. And who is to say that the scientific notions we give credence to today are definitive, and not distorted by tenacious, and mistaken, "obvious" conclusions?

I recommend this book to all those whose profession it is to train primary and secondary school teachers in the new Instituts de Formation des Maîtres or in the Centres d'Initiation à l'Enseignement supérieur. They will find themselves, as I have done, doubting their satisfying thought constructs and their reliable recipes; their outlook will change, and they will find themselves reworking their pedagogical strategies, putting to use the pertinent observations presented here.

I also recommend this book to those whose job it is to popularise science, communicate it or mediate it: this work shows how insidious and distorting certain images can be, even though they are a part of our vocabulary and everyday images. That common sense is double-edged is nowhere more evident than in language, and those whose job it is to use language, through necessity or choice, will exercise greater caution for having read this book.

I should also like to express the hope that work of this kind might be conducted more generally, and be taken up in other disciplines. Common sense, or, as the author calls it, *natural* thought, is common to us all, and is applied to all fields of knowledge: the type of research that is done on physics here could be developed further, but would be just as relevant in biology or chemistry, and colleagues who teach those disciplines or are interested in how they are taught will find this book thought-provoking and useful.

Pierre LENA,
Professor, Paris VII-Denis Diderot University

Introduction

Towards the middle of the twentieth century, two pioneers in education, Bachelard and Piaget, emphasised that knowledge, when presented, does not settle in empty or perfectly malleable minds which immediately adopt its forms.

Bachelard, in *La formation de l'esprit scientifique* (1938), develops the idea that all knowledge is built *against* what one already knows:

> In fact, one learns against previous knowledge, by destroying faulty knowledge, by surmounting what, in one's mind, is an obstacle to spiritualisation.

Common knowledge, according to Bachelard, presents characteristics which distinguish it clearly from the scientific approach. Its status being that of the evident, it is not open to refutation; but common thought is formulated in vague terms and is constituted of scattered and unrelated elements: it is knowledge in bits and pieces. To attain another – scientific – level of thought, one needs to surmount obstacles of a different nature. The substantialist obstacle, for example, consists in attributing a material nature to certain physical quantities, heat being a typical example.

This charting of stumbling blocks is not Piaget's main purpose. He attempts to characterise the development of intelligence through the successive capacities that emerge in children. On the basis of interviews with children or teenagers, he characterizes the types of intellectual processes that are or are not accessible to the interviewee, and goes on to determine what structures are available or not to the intellect. Without a particular structure, there is no hope of solving a particular type of problem. Thus, the subject's reaction, the way he or she copes with new knowledge, should, according to

Piaget, be seen as an indicator of the threshold of development he or she has or has not reached, which is a necessary condition of understanding. But a capacity is not a sufficient condition. Piaget's most significant and least controversial contribution is to make the subject's involvement a decisive factor in the learning process. And in his model of intellectual work there is the idea of a struggle with oneself, that was already present in Bachelard's epistemology. From the simple "assimilation" of new knowledge into an existing structure, to the extension constituted by "adaptation", which is in itself the result of a process of "equilibration", learning is always a question of negotiating with one's own knowledge (Inhelder and Piaget 1955; Piaget, 1975). Though of a different nature, Norman and Rumelhart's theory of information processing (1978) distinguishes between similar categories: "accretion", "tuning" and "restructuring" are reformulations, in terms that are very close to Piaget's, of the various forms such a negotiation may take, even though the scale of the modifications considered is very different in the two theories.

These authors do, at least, share the idea that knowledge is built both "with" and "against" what one already knows. This principle underlies the present study, and has, in the past twenty years, been decisive in inspiring a considerable body of work on the conceptions inspired by common sense (Johsua and Dupin, 1993).

Although there are a great many hypotheses on how such construction takes place, and on the ways to orientate it, this widely shared minimal position inevitably leads to one conclusion: it is preferable, when defining what is to be taught, to know the *a priori* ideas and ways of thinking of those one intends to teach. And if pupils are to take an effective interest in the knowledge that teachers are intent on conveying, they must be made aware that physics makes possible another kind of expression and activity, in a mode that is not that of natural thought. Paradoxically, if physics is to mean something, one must realise that it is often removed from common sense.

A good knowledge of the two aspects opposed here is therefore crucial: accepted theory on the one hand, familiar reasoning on the other, that is, the essentials of physics in contrast to natural reasoning.[1]

The forms of reasoning one adopts are not merely the product of chance. Recognisable trends of thought that are not compatible with taught theory are to be found everywhere, and are remarkably frequent and stable both during and after instruction, even in "higher" education. Numerous studies

[1] This does not apply to pupils alone. As Philippe Roqueplo (1974) has deplored, "Generally speaking, popularisers have a very vague idea of the readers they are supposed to address...; given these conditions..., what can they do but produce the best work possible and then cast it off... like a bottle into the sea?"

conducted worldwide on this subject concur. We must acknowledge the existence of these trends and realise their importance.

To stress that such reasoning is independent of any instruction received at school, the earliest descriptions referred to "spontaneous" or "natural" forms of reasoning. Some are manifest before any instruction in physics at school, and are therefore called "preconceptions". In some cases, at least, one would be justified in thinking that ordinary language and everyday experience are largely responsible for the convictions observed.

There are common lines of reasoning to which we are all attached. Their relative degree of coherence contributes to their resistance. If somebody were to come along and tell us they were erroneous, we would not give them up overnight.

But who *would* come along and tell us? Teachers? Yes and no.

They often do so indirectly, because what they teach does not as a rule contain "errors", or, more precisely, elements that contradict the established corpus: the knowledge they offer is coherent.

Nevertheless, this is not sufficient to cast light on, or to provoke a critical examination of, the trends of natural thought. Academic knowledge and natural reasoning may exist side by side in their individual territories. The result is considerable boredom in the process of learning and uncertain mastery in the end.

Familiar ways of thinking therefore deserve our particular attention.

This book is based on surveys involving students at various stages of education, from secondary school to university. It deals with certain basic elements of physics. Although it is primarily concerned with introductory lessons, many of these are central to the understanding of physics. These elements, then, are *foundations*, even though they were not immediately perceived as such in the history of ideas. Even if they cannot all be gone into explicitly at the beginning of the learning process, these points must be understood, at some stage, if one is to truly master a little physics, and, beyond that, a little science.

Research has shown that it is necessary to go over such points: these essential elements, though "elementary", are much less immediately accessible than they seem. The object of this text is to shed light on what makes them so difficult by taking into account common trends of thought.

Whoever wishes to arrive at a coherent conception of physical phenomena, or to inspire the desire for it in others, will have to take these trends into consideration.

This text is not exhaustive: not all the findings concerning common ways of thinking in physics are included here, even in summarised form. It is even less complete as regards the various types of research currently being conducted in science teaching. The work by Johsua and Dupin (1993) cited earlier, or the book edited by Tiberghien, Jossem, and Barojas (1998), among others, will prove useful for a further study of the aspects broached here and will provide readers with a broader view of this field of research[2].

In this book, the aim above all is to identify and to illustrate the main lines that organise natural thought in physics, placing them in counterpoint with those that structure scientific knowledge. This is the objective of the first (and main) part, after which the reader who is pressed for time can go directly to the conclusion. The second part presents a few complementary studies on various subjects, involving learners at different educational levels; the results bear out the analyses proposed in the first part.

REFERENCES

Bachelard, G. 1938. *La formation de l'esprit scientifique*, Vrin, Paris.

Inhelder, B. and Piaget,J. 1955. *De la logique de l'enfant à la logique de l'adolescent*, PUF, Paris.

Johsua, S., Dupin, J.J. 1993. *Introduction à la didactique des sciences et des mathématiques*, P.U.F., Paris.

Piaget, J. 1975. L'équilibration des structures cognitives, problème central de développement. *Etudes d'épistémologie génétique XXXIII*, P.U.F., Paris.

Roqueplo, P. 1974. *Le partage du savoir*. Seuil, Paris, p 31.

Rumelhart, R.D. and Norman, D.A. 1978. Accretion, tuning and restructuring: three modes of learning. In *Semantic Factors in Cognition*, J.W.Cotton and R.Klatzky, Lawrence Erlbaum Associates: Hillsdale, NJ.

Tiberghien, A.; Jossem, E.L. and Barojas, J. (Eds). 1998. *Connecting Research in Physics Education with Teacher Education*, HYPERLINK http://www.physics.ohio-state.edu/-jossem/ICPE/BOOKS.html.

[2] Additional references: This book was first published in 1996, in French. This English translation, written four years later, contains some additional references. As it aims at presenting some structuring ideas rather than a review of work on the subject, this book is still far from giving a complete account of all the research studies that are important in this field.

Part one

The main lines

Chapter 1

Physics: what is essential, what is natural?

How does the average person's approach to physics differ from the scientist's? First, we need to characterise physicists' physics and explain how we analyse the average person's reasoning.

1. THE ESSENTIAL: ABSTRACTION AND COHERENCE

Physics deals with constructs. It is true, of course, that falling bodies, the alternation of day and night or a river's flow are all natural phenomena and also objects of study in physics. But this does not mean that nature directly suggests what one should study in these phenomena in order to understand them.

The definition of physical quantities currently in use is the product of a lengthy process of abstraction. Energy, for instance, did not really make its appearance on the scene until the eighteenth century. The term *modelling* is often used to describe the correspondences established between reality and what one chooses to extract from it and to represent. This is done through measurements made with constructed devices; the information gathered is then fitted within, and checked against, theory. Scientific progress depends on complex adjustments between theory and findings, to better describe phenomena and forecast events.

The process always entails a simplification of reality. This can be achieved by thought. One can, for example, study the motion of a hammer without taking into account the action of air upon it. One can also simplify reality by "preparing" it. This is not easy to do for volcanic eruptions or supernovae, but laboratory physics is all "prepared reality", in which real situations are staged and controlled.

Simplification may not be the word that comes to mind before a jungle of computer-monitored apparatus, spitting wires and tubes in all directions. But that is the idea: to give an account of the complexity of physical phenomena, using as few quantities and relations as possible. Though applying this method indiscriminately to other complexities may not yield satisfactory results, in physics, at any rate, the method has proved a success.

In science, coherence is indispensable. A physical law cannot apply erratically. One therefore strives to attain the greatest degree of generality and to establish the extent to which the relations used are valid. Newton's theory of dynamics, for example, perfectly applies to velocities that are negligible in comparison with the speed of light. In terms of what is measurable, the theory applies well to the mechanics of ordinary objects.

Physics is based on rational simplification, abstraction and coherence. So how does natural thinking fit in?

2. COMMON WAYS OF THINKING IN PHYSICS

Determining the part played by natural thought in physics is an ambitious enterprise, and we have only partial answers. As we cannot photograph people's thoughts, we conduct surveys. But the fact that a question is asked, and the context in which it is asked, influences the answer. Reasoning always implies answering some sort of question. We have to accept the fact that the questions we put to people are not neutral, and take the questions into account when describing the act of "reasoning." As with quantum mechanics, the variations caused by the measuring instrument are part of the phenomenon observed.

And so we question people, in this case essentially pupils in the last years of secondary school, university students, and teachers, in more or less directive interviews or by using questionnaires. The questions can be closed, i.e., propose a limited number of potential responses; but that limits the scope of the investigation. At the beginning of a study, at least, it is essential to work from a much wider array of comments. The written or oral questions practically always come with a request to "explain" or "justify" the answer.

Classifying and interpreting these responses are difficult, overlapping tasks, and it is often necessary to provide a synthetic paraphrase of the statements collected.

Adding our own conjectures is dangerous: when we say, "the interviewee says this *as if* he/she thought that...", how are we to choose amongst all the possible interpretations? When interpreting comments made in a single questioning situation, we ought to keep to a near-paraphrase; otherwise what guarantee do we have? The information yielded is therefore very limited. If, however, a conjecture is tested over numerous surveys and is borne out, it proves more useful: on the one hand, its predictiveness increases, and on the other hand, it is easy to memorise and use, as it belongs to a necessarily limited set of "general" results.

The reliability and representativeness of the "findings" – interpreted facts – presented here varies.[1] They seek to avoid the two extremes: catalogues of paraphrased errors, or conjectures worded so vaguely that they cannot be refuted and so do not mean much. Between these two extremes, research in didactics has led to some noteworthy results.

The end product is a description of reasoning *trends*, that is to say modes of thought that are *likely* to arise in relation to a given problem. This should not be read as deterministic. "Natural reasoning" is used in the singular, but this does not mean that it is innate, or universal. Rather, the expression is used to describe relatively organised forms of reasoning that are widespread and tenacious, and that cannot be ascribed only to school learning.

3. TAKING "WRONG IDEAS" SERIOUSLY

So far, we have referred to "errors" and unorthodox reasoning in connection with common knowledge, although there is no reason for a study of common thought processes to focus first and foremost on their "incorrectness", or deviation from accepted and taught theory. That, however, is what first struck researchers in this field. Bachelard himself showed little interest in defining the "correct" aspects of common thought.

The simplest way, however, of determining that reasoning is rooted in common thought rather than in acquired knowledge is to check whether it is "right" or "wrong" according to accepted theory. When the reasoning is correct, credit goes to the school, when there has been any schooling, or to educated parents, who may have passed on their knowledge. Or else scientific "truth" is just held to be self-evident. When, however, the reasoning seems to contradict taught science, some questions arise. Error is therefore a good indicator of common knowledge.[2]

As if to make amends for focusing their attention on the interpretation of incorrect answers, most researchers in didactics have avoided negative connotations, describing them as "alternative frameworks", "notions", "common forms of reasoning"... Yet a major seminar was still, in 1991,

[1] Statistically speaking, percentages established for small test-groups are less representative.
[2] Closset and Viennot (1984) and all the initial studies considered it in this light.

entitled after one of the first terms used, "misconceptions"[3] – which proves that, beyond any value judgment, and although they themselves may not be aware of it, researchers are still primarily concerned with the ways in which common knowledge deviates from scientific knowledge.

Error is a valuable clue. And it shows the distance that remains to be covered on the road to learning. Of course, sustained relativism may lead one to deny the usefulness of such an approach, but, in the end, there are some things that all teachers would like students to understand,[4] at least the need for coherence, which we stressed earlier. That is what is truly essential, in physics.

Perhaps, however, frequent "errors" also have logic of their own that contributes to explaining their resistance. Investigating their possible "alternative coherence" entails providing a succinct description of natural reasoning, which would be anything but a simple catalogue of errors. The aim is to determine a few rules that can account for "average persons'" physics, just as the laws of physics do for scientific knowledge. This leads us to the question of how to define the various areas of study necessary to make the description more effective.

4. AREAS OF PHYSICS AND UNITS OF COMMON KNOWLEDGE: DO THEY COINCIDE?

Initially, research was deliberately orientated so as to focus on the specific contents of each discipline. This choice stemmed largely from a common consciousness of deficiencies in teaching, and was consistent with the idea that common knowledge is made of bits and pieces. At first, the fragments were patiently inventoried under headings that corresponded to chapters in textbooks. But later, a concern for synthetic description brought another configuration to light. The most strategic angles from which to analyse common knowledge do not always coincide with the stages of orthodox expositions, and adopting a transverse approach proves more fruitful than working from the accepted divisions of the subject. This approach makes it possible to organise findings in a simple fashion, and to orientate the search for new experimental facts.

[3] International Seminar on Misconceptions and Educational Strategies in Science and Education in Science and Mathematics, Cornell University (1983, 1987, 1991).

[4] This is the consensus that we describe as "correct answer" in the parts that follow, especially in the appendices and boxes in which survey questions are presented. The adjective "correct" might shock an epistemologist in that it implies selectiveness. Here it means, in short, "an answer that is accepted by the scientific community, within the context of a theory." We also use the term "physicists' answer".

The evidence thus compiled should prove that, if common knowledge is indeed made up of bits and pieces,[5] they are rather large; zones of coherence have, in fact, emerged from what first seemed a muddle of unrelated errors.

Now, on to teaching...

5. WHAT TO DO IN TEACHING?

5.1 Natural reasoning and teaching goals

Once the terrain has been mapped out, if one still wishes to teach a little physics, one must act. What "must be done" is not always apparent, and we must be wary of "all-it-takes-is…" formulae. However, let us briefly go over a few undisputable points.

First, a good knowledge of common ideas enables us to ascertain when they are unknowingly alluded to in certain textbooks: there are many spectacular examples of this. This increased vigilance is a good thing, especially when the authors of textbooks apply it themselves. Media popularisations, too, quite often bear the imprint of common modes of thought. There is a growing recognition of the pedagogical value of analysing such texts, and this type of research should contribute to such efforts.

Next, although many commonly held ideas may indeed hinder understanding, one should not rule out that some might actually constitute helpful bases on which to build knowledge.

Finally, when we seek to determine what we want to make our students understand, the results of this work are thought provoking. They can inform our choices and allow us to work on precisely identified difficulties. The tools used in the surveys – principally well-chosen questions – are also useful teaching tools. In any case, they should help dispel the illusion that by giving standard solutions to standard exercises, one has completely explained a problem. In physics, any subject broached is profound. There is no need to split hairs to prove this; one need only provide the proper illumination, although how to do this is not always obvious.

That is where defining the zones of coherence in common knowledge comes in – even if the coherence is erroneous. They may coincide with some elements of physics which, though not new, have not been explicitly taken into account in teaching. Indeed, they do not easily fall into traditional

[5] Bachelard (1938), like Moscovici (1976) in his study of the "social representation" of the object of a scientific theory, holds that common thinking is fragmented and insists on its lack of coherence. See also Di Sessa (1988): "Knowledge in Pieces".

textbook divisions. Does any secondary school textbook include a chapter on "the evolution of multi-variable systems," or "+ and - in physics," or "the time variable in the writing of physical laws"? Yet these are points that deserve our attention, as we shall see.

Organising experimental data on common reasoning makes it possible to isolate "sensitive" areas of elementary physics, where something might just fall into place for the learner. And so we must reconsider our teaching goals and ask: how do we define what is "essential"?

5.2 Legitimating pedagogical solutions

The pedagogical implications of the results obtained will generally be presented in the form of considerations and suggestions. The reader should feel free to diverge from the author's views on teaching, wherever they are not directly linked to the data provided – since in this field, conclusions can only be arrived at through experimentation, which is difficult. Trying out strategies at random would be easy enough. But to come to any useful conclusion requires a considerable research operation; a great many variables must be taken into account; circumspection, too, is needed. Work of this kind is now being accomplished, and will further the progress made in the past twenty years in the field that this book deals with.

This explains why the teaching suggestions presented here, when given in detail, are set in boxes or placed in appendices. This is not because the author believes that pedagogical action is less important than fundamental reflection, but to avoid a confusion of genres. On the one hand, some of the experimental facts interpreted here do lead to a description of what now prove to be unquestionably widespread reasoning tendencies. On the other hand, some pedagogical suggestions, though largely founded on prior analysis, are wagers of sorts; it has not yet been determined to what extent they are valid. This distinction is important.

The value of these suggestions is not, however, insignificant. They show that fundamental reflection on "what is essential and what is natural" in physics is not pointless, as it can lead to teaching goals and strategies that are precise and different from those that have traditionally prevailed. Moreover, these new proposals are sufficiently grounded for some of them to have gained the support of the groups of experts in charge of establishing the official syllabus and accompanying texts in France, between 1990 and 1995.[6]

[6] Technical groups made up of experts in each discipline (Groupes Techniques Disciplinaires, or GTD's) were created in 1990 on the initiative of Lionel Jospin. The members of the Physics GTD were, from its origins to 1995: Louis Boyer (President), André Calas, Hubert Gié, Jean-François Le Bourhis, Marie Thérèse Saglio, Jacqueline Tinnes, Laurence Viennot and Jean Winther.

Of all the considerations to be borne in mind when evaluating these proposals, one is of prime importance: will they obtain the support of teachers at large, in addition to that of their institutional representatives? Studies are under way[7], but a great deal of research remains to be done on this topic. This book aspires to contribute to the choices of those without whom all that is said about teaching is worthless – the teachers themselves.

REFERENCES

Bachelard, G. 1938. *La formation de l'esprit scientifique*, Vrin, Paris.

Closset, J.L. and Viennot, L. 1984. Contribution du raisonnement naturel en physique. in Schiele, B. and Belisle, C. (Eds.): *Les représentations. Communication -Information* 6 (2-3), pp 399-420.

Di Sessa, A. 1988: Knowledge in pieces. In Formann, G. and Pufall, P. (Eds.) *Constructivism in the Computer Age*, pp 49-70. Lawrence Erlbaum Associates, Hillside, NJ.

Moscovici, S. 1976. *La psychanalyse, son image et son public*. PUF, Paris.

[7] A European project, "Science Teacher Training in an Information Society" (DGXII n°SOE2CT972020, Coordinator R. Pinto, Group Leaders: J. Ogborn (UK), R. Pinto (Sp.), A. Quale (No.), E. Sassi (It.), L. Viennot (Fr)) is centred on this point. See Hirn, C. and Viennot, L. 2000. Transformation of Didactic Intentions by Teachers: the Case of Geometrical Optics in Grade 8 in France, *International Journal of Science Education*, 22, 4, pp 357-384.

Chapter 2

A trend in reasoning: materialising the objects of physics

Examples from elementary optics

In association with Françoise Chauvet and Wanda Kaminski.

1. THE ESSENTIAL IN PHYSICS: CONSTRUCTED CONCEPTS

The physical sciences describe phenomena in terms of physical quantities and laws. In this process of abstraction, where concepts are constructed, familiar notions are of little use. The process is, in fact, far from natural, as the entire history of science proves.

This difficulty is particularly apparent in the field of elementary optics; the theory can be approached relatively simply, and the essential laws can be summed up in a few words. But even at an elementary level, optics involves constructed, and therefore abstract, concepts. A “ray of light”, for example, is not a material object. It is a mode of representation, often called a “model”, used to translate into symbolic language the propagation of light. Thus a ray does not have the same status as ordinary objects, such as a table or a chair. In particular, we cannot see rays of light, and those we think we can see, be they “rays of sunlight” or “laser beams,” are in fact diffusing particles, each illuminated by a rectilinear narrow beam. Although an “optical image” can indeed be seen, the rules of its formation are surprising, and very different from those governing ordinary material objects. How does natural reasoning approach these immaterial objects? Is their difference from ordinary material entities grasped? From the same perspective, is it easy to accept that the concept of colour must be distinguished from our idea of what characterises an object? These questions are not as naive as they seem, as we shall see.

2. COMMON FORMS OF REASONING IN ELEMENTARY OPTICS

2.1 Light and vision

2.1.1 What needs to be understood

First of all, the propagation of light can be described by means of the model of rays of light: the path of light is then imagined as a line in space.

Box 1

Light, the eye, and how we see objects, according to pupils aged 13-14, before instruction on the subject (Guesne, 1984)

Question:

How do we see an object?

Answers:

The answers can be summed up in four diagrams

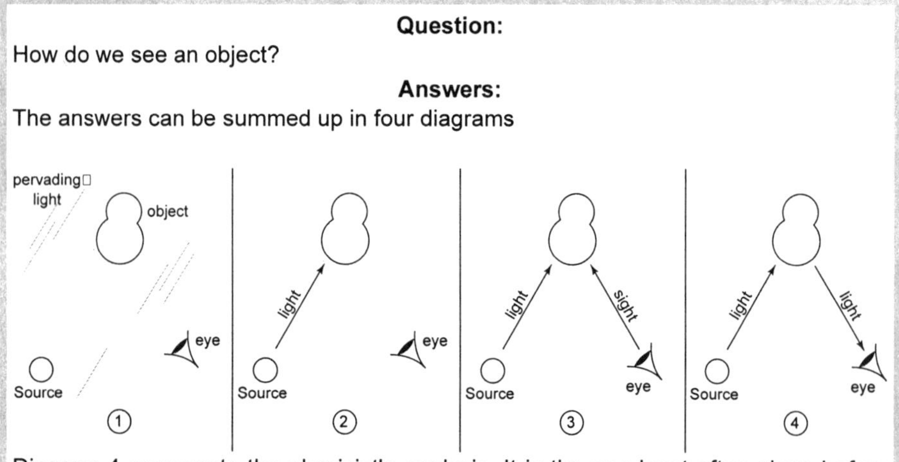

Diagram 4 represents the physicist's analysis. It is the one least often given before instruction.

Diagrams 1, 2, and 3 represent the most frequent comments, such as:

DIAGRAM 1: "The eyes need daylight to see clearly, they need light" (no further details – age: 14).

DIAGRAM 2: "It's thanks to light that you can see the box" (no further details – age: 14).

DIAGRAM 3: "This is how the eye sees, it goes like this (drawing of lines emanating from the eye). The eyes have no light of their own, so they need light to illuminate what you want to see" (age: 13).

Barring accidents, such as a change in medium or a non-homogeneous medium, its path is a straight line.[1]

Secondly, for vision to take place, the light emanating from the object must enter the eye.

These laws seem simple. But they are not naturally applied by children,[2] who more readily identify light with its source, with the illuminated surfaces that they observe, or with a kind of pervasive glow, than with an entity which conveys information to the eye (box 1). Rectilinear propagation and light entering the eye are not tools that they use as a matter of course in their reasoning.

What is the situation just before or after students graduate from high school?

2.1.2 Question: Punched screens

From Kaminski (1989, 1991); see also Chauvet (1990).

Adults in technical vocational training (Applied Arts section) and middle-school teacher trainees were asked to answer the following question (box 2):

> What can one see from each of the holes H1, H2, H3 looking through the small hole H, when the bulb lights up?
>
> Explain using diagrams.

The correct answer is: "From hole H3 one will see the lit bulb, and from the two other holes the black screen," or, less precisely, "Light from hole H3 and black from the others". The problem can be solved by using only the two laws stated above; the explanatory diagram is:

[1] It is necessary to adopt another model, that of waves, when considering spatial dimensions closer in size to the wavelength of the considered wave, but this is not the case in the examples that follow.

[2] Guesne et al. (1978), Guesne (1984) and Tiberghien (1984a).

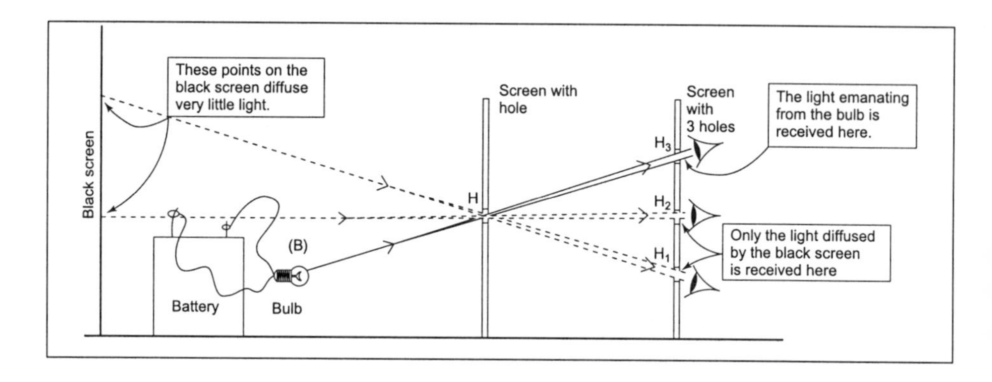

Because the holes are several millimetres in diameter (or ten thousand times as big as a wavelength of the visible spectrum), it is not necessary to question the rectilinear propagation of light, as the diffraction is not observable.

The results are summarised in box 2. In the case of hole H3, which is "opposite" the lamp, all the participants predict that there will be an impression of light, but no more than a quarter of those interviewed say that it will be possible to see the lamp itself. As regards the other holes, at least half those interviewed (50% of the teachers and 65% of the students) predict (wrongly) that "hole H would be bright," or that "there would be light." Some diagrams associated with erroneous predictions show lines that diverge from the first hole (H), but only in half of the cases do these lines reach the eye. And some comments specify that these are lines of sight and not paths of light.

These results indicate that the rectilinear propagation of light and the need for light to enter the eye are not constraining laws in the reasoning of educated adults. A small proportion adhere to them rigorously, but most people reason as if light were itself an object, visible from just about anywhere.

Box 2
Question: Punched screens

Question

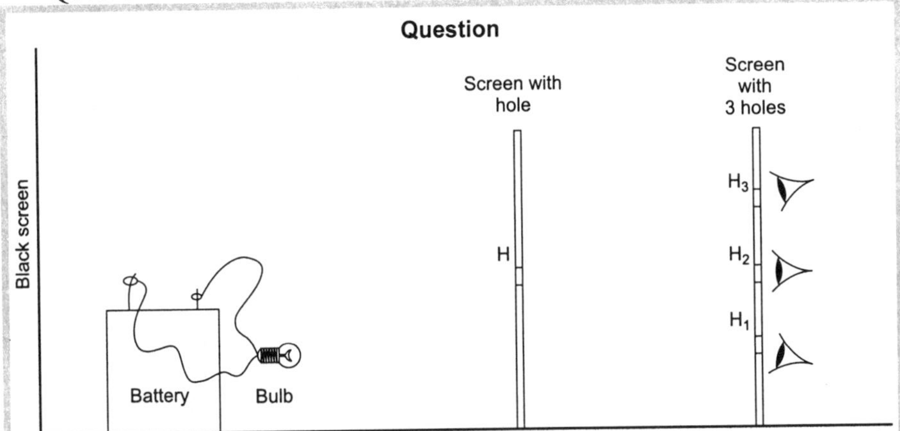

What can one see from each of the holes H1, H2, H3 looking through the small hole H, when the bulb lights up?
Explain using diagrams.

Results

One can see...	Trainee teachers (n = 36)		Students of Applied Arts (n = 48)	
	Through H3	Through H1 and/or H2	Through H3	Through H1 and/or H2
1. something bright (luminous spot, light, luminous hole, beam, source, bulb, ray...)	100%	50%	100%	65%
specifically: bulb or source	25%	-	15%	-
2. a black hole or nothing	-	25%	-	30%
3. the black screen		0%	-	5%

The diagrams obtained are of three types (study of trainee teachers)

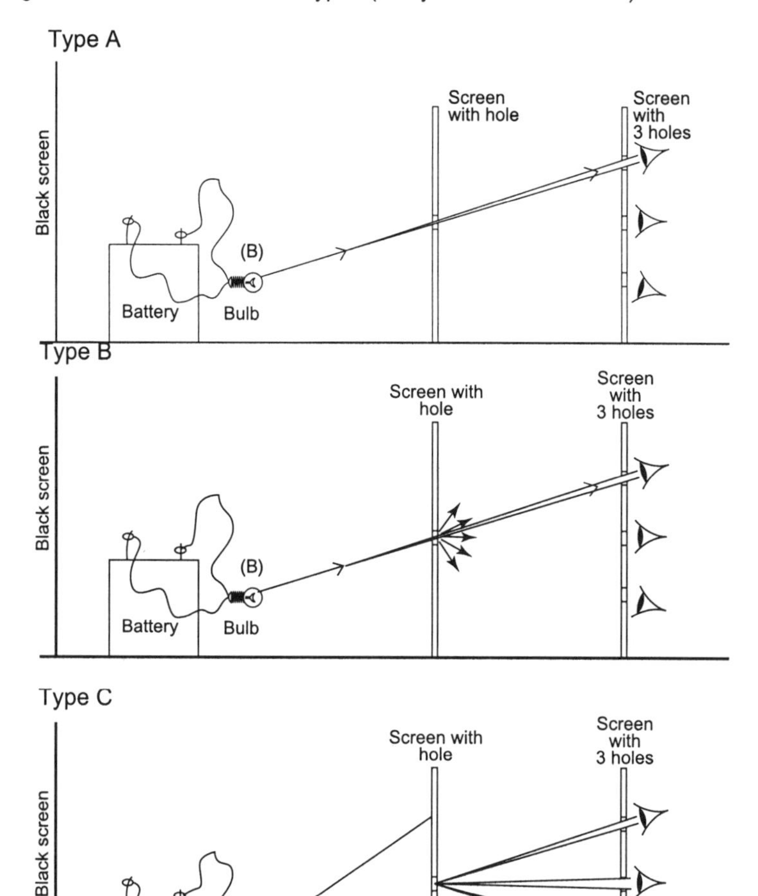

The diagrams provided which contain obvious errors (answer 1 for H1 and/or H2) are of the B or C type.

Notes:

On some B-type diagrams, no line reaches the eye.

Some comments accompanying C-type diagrams say that the lines represent lines of sight and not paths of light.

2.2 Optical images

2.2.1 Basic notions about optical imaging

The concept of an optical image is complex. We approach it here within the framework of the model of rays of light, and in its simplest form: any ray of light emanating from a point of an object and traversing an optical system "passes" through another point on its way out – the image point (box 3). The term "passes" is a simplification: the straight line representing the ray passes through the image point, even though light in the real sense of the word may not (it is sometimes said that the ray or its projection passes through the image point). It seems simple, and in fact this definition is rarely developed further in the classroom. On the other hand, it is very often used in geometrical constructions which make it possible to locate the image of an object in a given optical system: from two known rays emanating from one point, one can predict the path of all the other rays emanating from the same point; those two rays are enough to locate the image. The famed "construction rays" are those whose path it is easy to trace. Box 3 shows the prototypical diagram of the construction of the image of an object in a converging thin lens.

This technique is based on sampling. Only a few points of an extended object are used to predict the position of its complete image, and for a given object point, only two rays are used to predict the path of all the other rays emanating from the same point and interacting with the optical system. Of course, it is difficult to "count" the rays and the "points" of the object. But the principle is to analyse the continuous by means of the discontinuous.

The models of the rays of light and of the formation of the focussed image can, at any rate, illustrate two facts:

- to form an optical image, there must be an optical system or a non-homogeneous medium. Otherwise, the rays emanating from an object point no longer cross, but diverge from that point in all directions.
- a small part of a thin lens is enough to form the image of each object point, and therefore the entire image of an object. Only the brightness of the image is affected by a reduction in the effective surface of the lens.

In other words, the information provided by light emanating from a given point of the object is completely "spread out" in space unless an optical system reassembles it somewhere. And this spreading out makes it possible to retrieve the information with a limited part of the lens, provided the light carrying the information reaches it.

Box 3

Optical image: point to point correspondence between object and image; case of a converging thin lens

Object point

Image point

- Any ray emanating from the object point and hitting the lens passes through the image point.
- The area in grey shows the progress of a beam of light. It provides enough information to form a complete, though less luminous image.

Some rays are easy to construct. For example, those that pass through the centre of the lens are not deviated; those that are parallel to the axis pass through the "back focal point" as they leave.
This leads to the typical diagram for the construction of an image (A'B') of an object (AB) in a thin lens (with focus F and F').

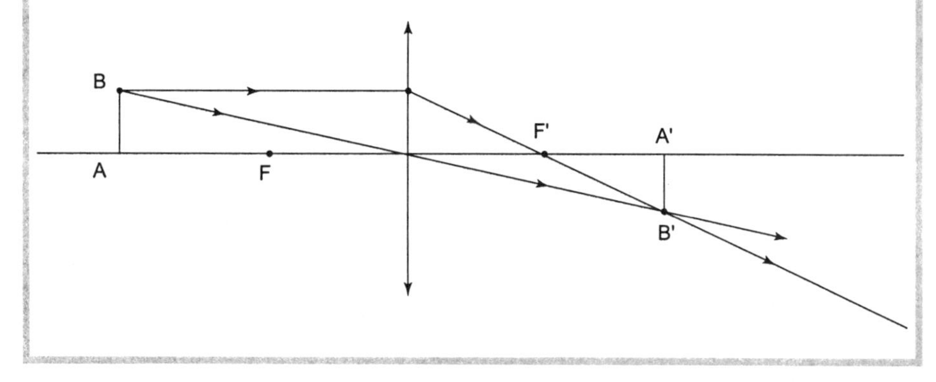

So, this apparently simple explanation of the focussed image introduces pupils to quite a bit of physics – and it is conveyed through words and diagrams exclusively.

Box 4

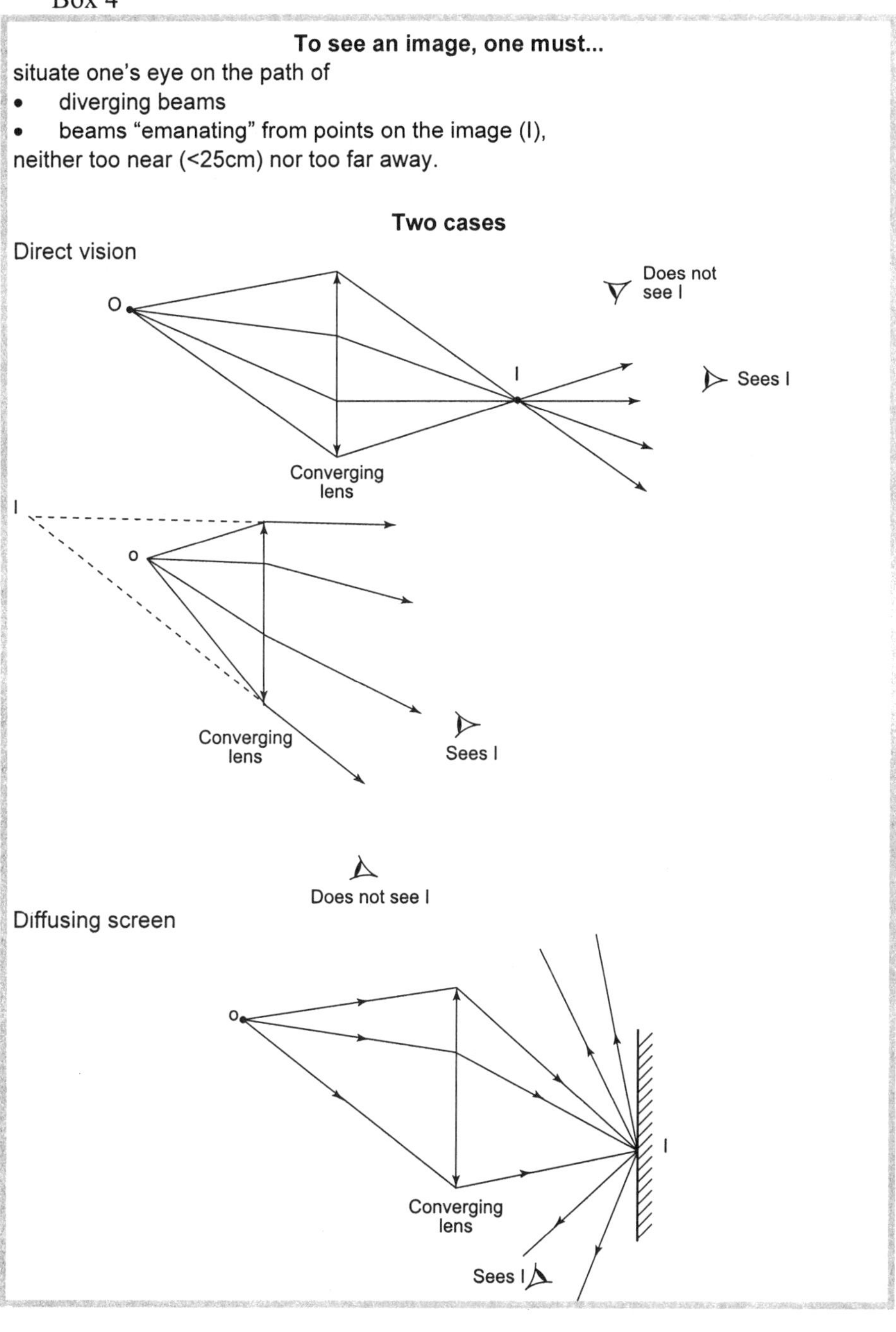
To see an image, one must...
situate one's eye on the path of
• diverging beams
• beams "emanating" from points on the image (I),
neither too near (<25cm) nor too far away.
Two cases
Direct vision
O
I
Does not see I
Sees I
Converging lens
I
o
Converging lens
Sees I
Does not see I
Diffusing screen
o
I
Converging lens
Sees I

The model of the ray of light and the explanation of the focussed image make it possible to analyse[3] much of what can be achieved and observed as regards optical systems and images.

But how do we see images?

With the eye, provided that the optical image behaves like any other visible object: from each of its points a diverging beam of light must enter the eye; the image must be properly placed, neither too near nor too far from the eye, so that the transparent parts of the eye can form a new image on the retina from this light. It is therefore important that, in direct vision, the eye be in the path of the beams. If the luminosity is adequate, a diffusing screen makes observation easy, as the light leaves every point of the image in a very wide range of directions (box 4).

Of course, all sorts of instruments can form images which are not optical images, based on other principles. And receptors other than the eye can be used to analyse optical images.

But we have chosen to focus on a basic principle to see how it is understood.

Two seemingly simple survey questions have yielded surprising results.

2.2.2 Question: The removed lens

This question was proposed to first-year university students in the United States, to students specialising in Optics and to pupils in grades 11 and 12 (science section)[4] in France, and to pupils in grade 11 in Lebanon.[5]

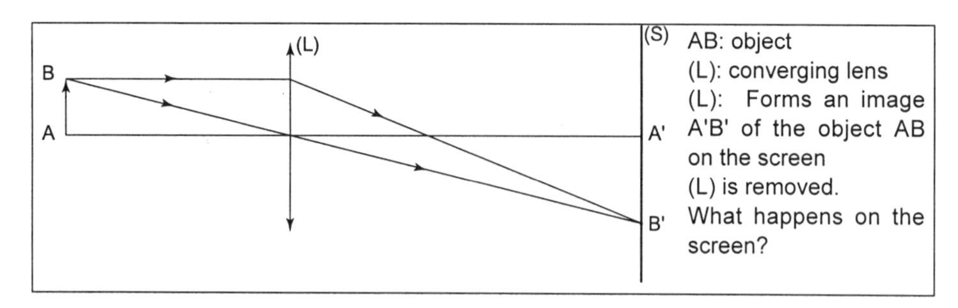

Another version of the question does not mention the existence of the lens at the outset. Without the lens, the question is simply: "What does one see on the screen?"

[3] Like all models, it "depicts" reality – not absolutely, but within a margin of variation which the nature of the measurements allows.

[4] Section de Techniciens Supérieurs d'Optique, Première et Terminale scientifique.

[5] I.e., Première.

Without the lens, the screen is almost uniformly illuminated: the light emanating from each point of the source is "spread out" in space before reaching the screen, and overlaps with that emitted by the nearby points.

Box 5
The removed lens

Question

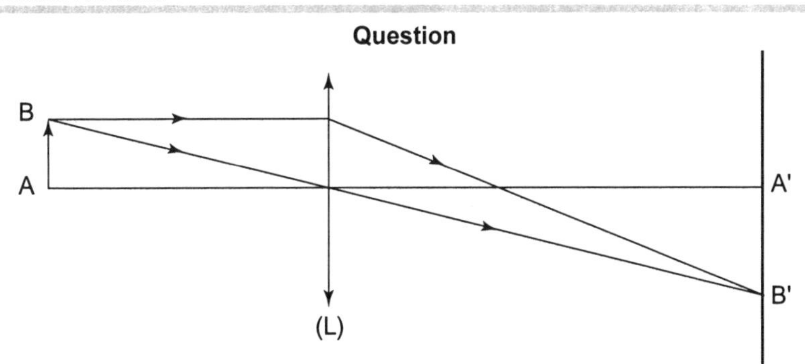

(L) is removed. What happens on the screen?

Correct answer

Without the lens, the screen is almost uniformly illuminated.

Results

Resultats	American students [a]	Pupils in Lebanon [b]	Paris [c]		
			Optics students[1]	Pupils in grade 12 (math and science sections)[2]	Pupils in grade 11[3]
No image	50%	45%	55%	30%	40%
Image specifically: erect image	40% 40%	45% 40%	40% 40%	55% 50%	40% 40%
Other, or no answer	10%	10%	5%	15%	20%
Number of participants	22	31	54	73	24

1) Brevet Technicien Supérieur d'Optique.
2) Terminales C et D
3) Première F
a. Goldberg, F.M. and Mac Dermott, L., 1987. An investigation of students' understanding of the real image formed by a converging lens or concave mirror, American Journal of Physics, 55, 2, pp 108-119.
b. Fawaz, A. et Viennot L., 1986, Image optique et vision, Bulletin de l'Union des Physiciens, 686, pp 1125-1146.
c. Kaminski W., 1986. Statut du schéma par rapport à la réalité physique, un exemple en optique. Mémoire de Tutorat. DEA de Didactique, Université Paris 7.

Over 40% of the students believe that an image can be formed even without a lens. Almost all of these students state that the image is then erect:

"AB is not deformed because the luminous beam only crosses air." (Student – grade 12)

"Without a lens, A'B' is not inverted and it is the same size as the object." (Student, grade 12)

Diagrams like the one below illustrate this idea:

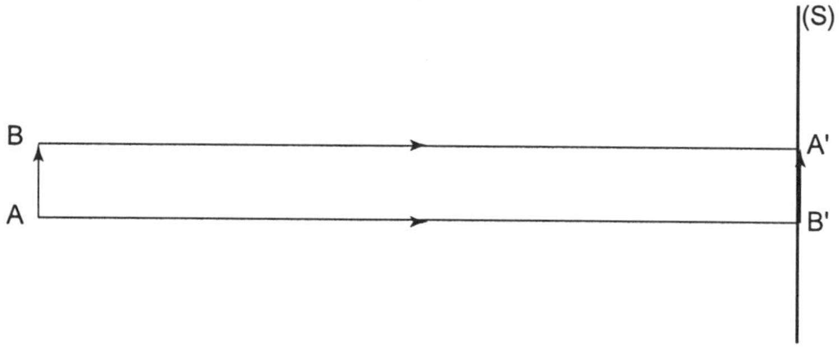

(S)

B A'

A B'

A test question is asked to determine the effect of mentioning the lens at the outset.

(S)

Question :

AB: object

E: screen

What does one see on the screen?

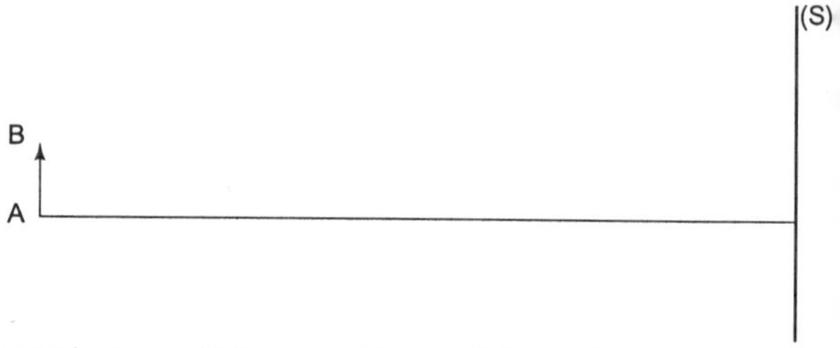

B

A

In grade 11 in Lebanon, 46% of the students (n=28) predict there will be an image.

The results given in box 5 are astounding. And they are similar for both versions of the question. After sometimes substantial instruction in optics (for the Optics students, for instance), describing the means by which images are formed, many students (between 40% and 55 %) state without hesitation that one can do without such means: according to some, the image of the source makes its own way towards the screen represented in the proposed diagram. "There is no longer an image-deforming system," comments one student in order to justify a fact that many others predict: that without a lens, the image that appears on the screen is no longer inverted. The diagram provided to justify one answer, reproduced in box 5, shows that travelling has not affected the size of the image. According to this model of global transportation,[6] it is practically a moving object. There is no question of diluted information which is then concentrated elsewhere.

[6] See the "holistic model" of Feher and Rice (1987), the "travelling image" of MacDermott (1987), Fawaz (1985), and Kaminski (1989). See also Galili (1996), Galili and Hazan (2000).

One can wonder at this: how could the image know in which direction to go? Can it guess where the screen is? Nothing in everyday experience supports such ideas[7] (has anyone ever seen the image of the filament of a bulb on an adjacent wall?), so where can they come from?

It seems, at any rate, that the need to think in terms of an identified, concentrated object, prevailed with many students.

Box 6
Masked lens

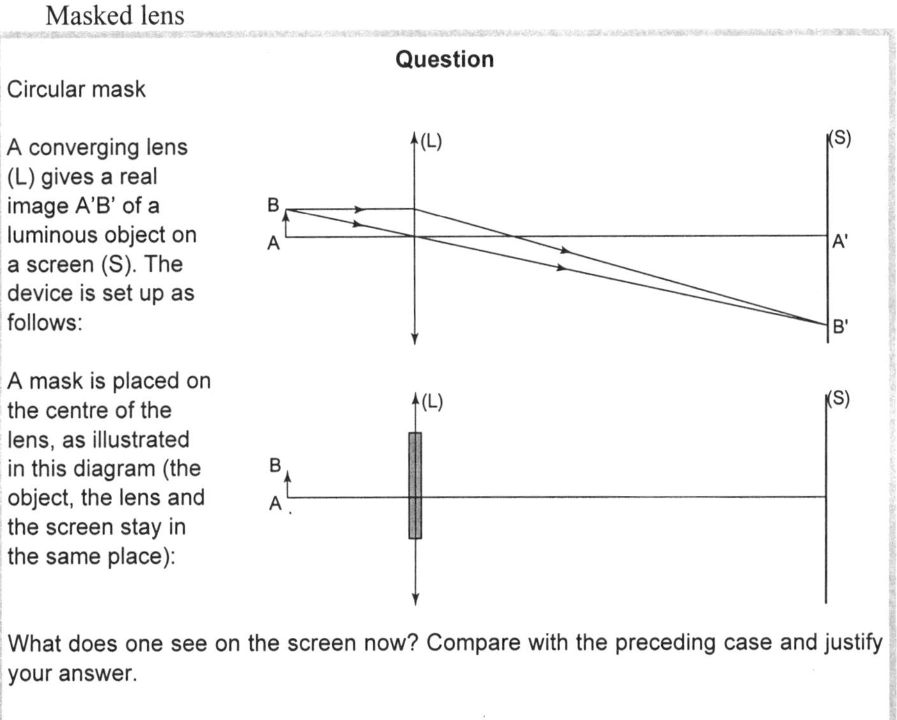

[7] It is very commonly (not to say "universally") allowed, among researchers in didactics (see also Bachelard, 1938, 1966), that common knowledge is rooted in everyday experience. That is more than likely, but one must point out, in the case of the "travelling image," that "naive" ideas do not always stem *directly* from everyday life.

Correct answer

A limited part of the lens is enough to form the image (see box 3), which is then less bright (and not as clear in a case of imperfect focussing).

Results

Answer / Population	One sees the complete image	One no longer sees any image, or One sees a part of the image
Last year of secondary school in Lebanon (N=58)	27%	66%
First-year university students, Paris 7 (N=93)	42%	46%

There is no image, or an incomplete image, because:

"You can see the outer edges of the image because the centre is hidden."

"If the object is smaller or the same size as the mask, you cannot see any image."

"You will see the shadow of the mask on the screen."

"You see a luminous point emanating from B, and which can reach us from either the top or the bottom of the lens."

a. Fawaz A., Viennot L., 1986.

2.2.3 Question: Masks

If the image is understood as a moving object, an obstacle should punch a hole in it like a die. The situation proposed therefore involves placing a mask on the lens. Many versions are possible, but let us limit ourselves to one, reproduced in box 6: the mask is placed in the centre of the lens and the participants are asked how this affects the image of an object.

In the hypothesis of perfect point to point correspondence, the correct answer is that the mask affects only the brightness of the image. Indeed, as is shown in box 3, any portion of the lens can form the image of every object point. A complete and discriminating answer might cast doubt on the hypothesis of perfect focussing.

It is massively apparent from the answers (see box 6) that the students believe a mask placed on a converging thin lens creates a hole in the image that appears on the screen.

If one imagines the image as travelling,[8] reaching the lens, being punched as it passes the mask, being inverted as an image because it has crossed a lens, and moving on as far as the screen to settle there, then one will have reconstituted many of the students' answers, such as:

> "You can see the outer edges of the image, because the centre is hidden";
>
> "You will see the shadow of the mask on the screen";
>
> "You see a luminous point emanating from B, and which can reach us from either the top or the bottom".

So this is the "image" that arises from these common answers: a travelling, concentrated entity, which sometimes leaves fragments along its path.

2.2.4 Substantial objects

These first results show the distance separating physical theory from common trends of thought. Light as a carrier of information to the eye, the non-material (in the ordinary sense of the term) nature of rays of light, the dilution of luminous information in space, except at the object and image points – all these aspects are not naturally considered in our reasoning. On the other hand, the idea of a substantial object seems to underlie many answers: light is visible "from the side," in itself (without a diffusing object), an image is fully formed from the beginning and moves in one piece, possibly leaving some fragments along the way if an obstacle has punched a hole in it in passing.

As we shall see, this idea of an object, and the related idea of the characteristics of an object, also considerably affects how students ordinarily understand colour.

[8] The expression "travelling image" sums up a mode of understanding the concept of an image which may explain certain elements of the answers obtained for various questions (such as the two quoted here). But other aspects of the answers can probably be ascribed to other causes. This is especially the case for comments which involve elements learnt at school. "Construction rays" often appear to have undergone a change in status: these simple tools of geometrical construction sometimes end up being considered as real components of the image. For example, when the "central" ray cannot pass through the lens because of a mask in the centre of the lens, some students state that, because of this, the image is cancelled (Fawaz, 1985; Fawaz and Viennot, 1986).

2.3 Colour

Based upon Chauvet (1993, 1994, 1996).

2.3.1 What needs to be understood

Colour is a perceptual response of the eye to luminous stimulation. A light produces a sensation of colour if certain radiations (i.e., radiations of certain "wavelengths"[9](λ)) are missing from those that constitute white light.[10] When the radiations that are present have very similar wavelengths, it is said that the light is (almost) monochromatic.

A colour can be associated with any wavelength of the visible spectrum (with a value of λ between 400 and 700 nm – see box 7), but the opposite is not true. For example, when two laser beams, one red, one green, are projected on to a single part of a screen (a screen that is white when illuminated in white light), it will turn a most beautiful yellow, without the presence of any "yellow wavelength" in the spectrum of the light reaching the screen. This shows that colour cannot be reduced to the associated wavelength.

This effect is due to the structure of the receptors of the retina. Box 7 sums up the rules resulting from this structure.

When several coloured lights enter the eye simultaneously, one speaks of "additive mixing". The rules in box 7 make it possible to conduct a simplified analysis, based on lights that each make up a third of the visible spectrum, red, green and blue respectively. These are the primary colours for additive mixing. Thus the additive mixing of a light which is the red "third of the spectrum" and of another which is the green "third of the spectrum" produces yellow light. The receptors of the retina are such that the additive mixing of monochromatic lights which are respectively red and green[11] gives the same result: one sees the colour yellow in spite of the "hole" in the spectrum around the "yellow" wavelengths.

The colour of bodies illuminated by sources of light brings in another aspect: the action of matter on light. This often involves selective absorption, where matter diffusely reflects only a part of the radiations it receives. Materials are characterised by the band of radiations that they absorb, i.e.,

[9] Radiations are also characterised by their frequency $f = c/\lambda$, where c is the speed of light in vacuum, λ being the wavelength in vacuum. When they enter an environment with an "index of refraction" n, the speed and the wavelength of the light are divided by n, and the frequency remains the same.

[10] This is not a necessary condition: certain perceptual phenomena (simultaneous or successive contrasts) can make one see colour on surfaces illuminated with white light.

[11] Everything that is said here on additive syntheses presupposes that the lights are properly balanced in intensity.

that they do not diffusely reflect, or, in the case of filters, that they do not transmit. When we say that an object is yellow, for example, this essentially means that it absorbs blue, so that it appears yellow in white light. When illuminated with blue light, it appears a very dark grey. Therefore objects do not have one colour, but are transformers of light. When an object appears white, this is because it diffusely reflects the entire visible spectrum without notable alteration.

When several subtractive actions occur at once, for example when overlapping filters are placed in the path of light, or when pigments are mixed finely, there is a "subtractive mixing". The light that emerges is the light which was absorbed by neither of the materials. Box 7 gives the corresponding rules, on the basis of lights that are "thirds of the spectrum", as before.

The results obtained by superposing two beams of light of different colours are, therefore, very different from those obtained when mixing the same two colours in painting. To give a brief example:

- with lights, red + green makes yellow,
- with paints, red + green makes brown.

Lights and paints: how is this double phenomenology commonly apprehended?

2.3.2 Question: Red + green

The following question was proposed to students in Applied Arts[12] before a physics class on colour:

> During a show, a beam of red light and a beam of green light are projected on the same spot on a white backdrop. What does one observe at the spot where the beams overlap?
>
> white brown yellow red and green I do not know other

The correct answer is "yellow".

Only 3% of the students (N=60) gave the correct answer, whereas half of them answered "brown". This result testifies to their experience in mixing paints, which is also apparent in the following excerpt from an interview.

Two students of Applied Arts (A and B) and their teacher (E) are discussing the experiment where red and green lights are superposed, and contemplating the magnificent yellow colour which this produces:

[12] Techniciens Supérieurs d'Arts Appliqués.

Box 7

Additive and subtractive mixing: colour associated with "thirds of the spectrum"

Separating the various radiations that constitute "white" light gives a "spectrum".
The spectrum of white light ranges from λ=400nm to λ=700nm [a].

Additive mixing

Coloured lights with a spectrum corresponding to a third of the spectrum of white light

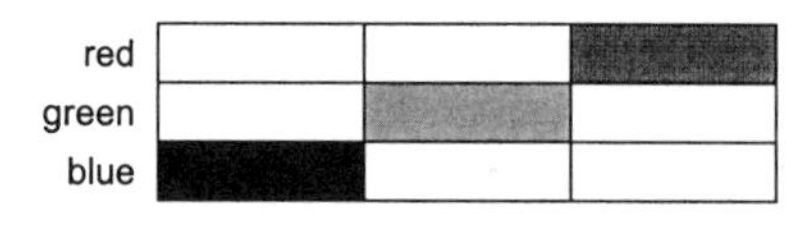

Combining two of these lights in the correct proportion respectively gives a light that is

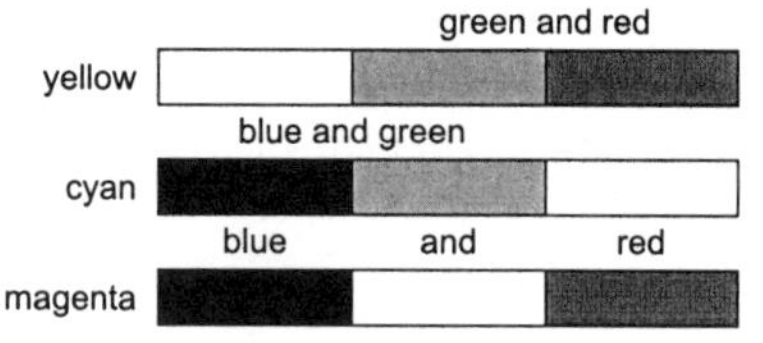

Subtractive mixing

A filter (or a pigment) absorbs a part of the spectrum of white light:

- A **yellow** filter absorbs **blue** light and diffusely reflects **green** and **red** lights.
- A **cyan** filter absorbs **red** light and diffusely reflects **blue** and **green** lights.
- A **magenta** filter absorbs **green** light and diffusely reflects **blue** and **red** lights.

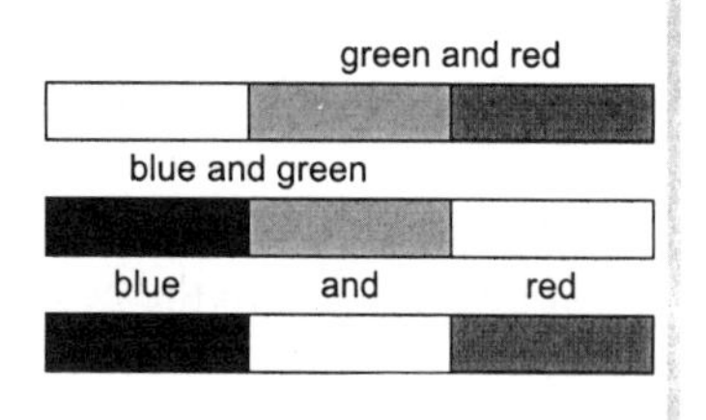

When illuminated in white light, two superposed filters or two blended pigments send back (transmit or diffusely reflect) to the eye only the part of the spectrum that they have in common:

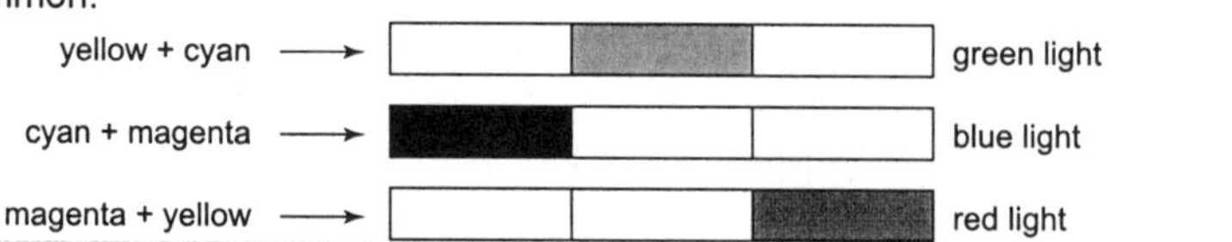

a. λ : wavelength; 1 nm= 10^{-9}m. The spectrum is here schematically divided into three thirds.

E: What do you see?

A: Something that is almost ochre... not ochre, but brownish; when you mix paints, in theory, mixing red and green generally gives a brown colour.

E: But when you look at it, is it brown?

B: Actually...

A: It's a shade of orange.

B: I see yellow.

This nicely illustrates the fact that when one knows "colour as matter", one does not readily accept the rules of the additive composition of coloured lights. This reluctance is so great that "experimental evidence" is recognized only after considerable effort.

These answers and comments show that the students think of colour as intrinsic, ingrained in matter, and obeying the rules that apply to paints. Light is thought to obey the same rules as objects. What follows is another example of this.

2.3.3 Question: Overlapping laser beams

> Two laser beams of different colours,[13] one red, one green, cross in space. Will the light of each be modified after crossing the area where they overlap?

A third of a group of students preparing a Diploma in Art Professions[14] (N=14) predicts that the beams will have a lasting effect on each other, i.e., that their colour will be modified beyond the zone where they overlap. When explaining their reasoning, the students seem to think that the coloured beams either dye or filter each other:

[13] By the "colour" of a beam we mean the colour one would observe when putting a white screen in its path. (This note is in the text of the question.)

[14] Diplôme de Métiers d'Arts.

"No, the colours will not be the same after crossing, there will be a mixture of colours";

"Each coloured beam acts as a filter for the other".

Coloured light is, in a sense, "reified".

3. CONCLUSION

An essential aspect of natural reasoning emerges here: the tendency to analyse phenomena in a reductionistic manner, a common material object serving as an implicit reference for the analysis of abstract concepts. Nevertheless, these examples from optics concern objects that are not material in the usual sense of the word: an optical image or a beam of light does not weigh much, and the sensation of colour still less.[15] But we have seen that this does not prevent a certain substantialisation of concepts from entering into reasoning.

Therefore, the idea that physical quantities are constructs, and not the products of a self-evident reality, is not a natural one. The material status commonly attributed to physical concepts can be linked to what Bachelard (1938) called "the substantialist obstacle."[16] The tendency to consider the objects studied in physics as material objects is manifest in many situations described in this book, and has serious consequences.

As far as instructional methods are concerned, each of the studies mentioned in this chapter provides valuable information for teaching optics.

Primarily, they all highlight the choice of objectives. Are the conceptual elements involved essential to a basic understanding of optical phenomena, or not?

If we decide that they are, the results reported here will help us choose specific strategies. We cannot simply assert that light travels in a straight line, illustrating this rapidly by "rays materialised by dust". Applying algorithms for the geometrical construction of images using "central" or "parallel" rays should be attempted only after considerable preliminary work on the principle of the formation of images itself. We should not allow

[15] Other objects in physics could have been studied in this chapter, particularly atoms and molecules, to which teenagers often attribute the (macroscopic) properties of the bodies they constitute (see Driver et al. 1985 and Lijnse et al., 1990).

[16] Indeed, the "substantialisation of an immediate quality" consists in claiming to explain a phenomenon by attributing to a substance an intrinsic property, which makes it responsible for the phenomenon: "The immediate phenomenon will be taken as the sign of a substantial property.... Thus one attributes to electric fluid the quality of being 'glutinous, smooth, resistant'.... 'Electricity is like glue'." Bachelard (1938, p 102-103).

colour to be identified with wavelength some of the time, and seen as an intrinsic property of matter at other times; we must define it as a perceptual response of the eye to light that has itself been transformed by matter. For each of these themes, our aims must be to emphasise what lends coherence to the concepts being taught and to work explicitly on what poses problems for our intuitions; the following points, in particular, should be insisted on:

Light is a carrier of information to the eye; it is not visible "from the side", only diffusing objects (dust, etc.) create this illusion; between the source and the eye, the luminous information undergoes geometrical alterations (spreading out, concentration) or transformations in spectral composition, through the action of matter; a few simple rules (point to point correspondence, additive and subtractive mixing) allow us to give a unified description of many phenomena involving images and colours.

Keeping these objectives in mind truly transforms the teaching of elementary optics, in spite of the apparent triviality of the content matter and of the experimental settings used. Everything depends on what the teacher makes of these settings: the types of activity, the questions and discussion that follow. To take just one example, the visualisation of a beam of light using a diffusing medium is often exhibited without preamble as "proof" of rectilinear propagation, and the fact that for vision to take place it is necessary for light to enter the eye is simply asserted. An alternative strategy, based on the analysis above, is to broach the second point after careful preparation (see appendix 1 for suggested activities), the visualisation of the beams being introduced later, to produce a synthesis of rectilinear propagation, diffusion and vision.

It is at any rate useful to be aware of some situations, such as those described in boxes 2, 5, and 6, in which observation contradicts natural reasoning. The surprise caused by unexpected effects, and the pleasure of making correct predictions afterwards, are powerful stimulants, which foster discussion in the classroom and make the teacher's comments more effective organising agents.

These principles and goals underlie the recent teaching proposals in France,[17] which are based on the didactical research and suggestions summed up here.

Let us return now to the analysis of common reasoning, and examine the consequences of the tendency to "reify" conceptual objects, a tendency which, as we have seen, is all too obvious in elementary optics.

[17] See appendix 1 on the convergences between the national syllabus at grade 8, or Quatrième, launched in 1993 in France, and findings in didactics; see appendix 2 for the accompanying document on the use of the pinhole camera as a device; and, for a sequence on colour, see Chauvet (1993, 1994).

APPENDIX 1

RESEARCH IN DIDACTICS AND THE NEW FRENCH SYLLABUS: CONVERGENCES.

The example of the French syllabus at grade 8 (1993)[18]

This appendix sums up, with a few minor modifications, an article published in *Didaskalia,* in French (Viennot 1994).

INTRODUCTION

In 1990, major changes were made in French secondary education. Technical Groups made up of experts in each discipline (Groupes Techniques Disciplinaires, or GTD's) were appointed for five years: their mission was to formulate proposals on the syllabus for each discipline, the global framework being defined by a National Council for the Syllabus (Conseil National des Programmes, or CNP). The two separate GTD's on Physics and Chemistry worked in association with each other. These two disciplines have, since 1990, been taught only in the last five years of secondary education, from "Quatrième" to "Terminale."[19] The proposals made for these five grades were implemented, after consultation and adaptation, between 1993 and 1995.

To what extent do the results of this effort toward renovation converge with the findings on teaching in each discipline – in other words, with research in didactics? First of all, in this case, what is called "didactical research" is not limited to research findings. There are consensual points of view in didactics that are not, strictly speaking, "results," and it is worth adopting a critical approach. To begin with, decisions on teaching objectives are dictated by politics rather than by research. Of course, the latter can be crucial, providing information on accessibility, determining conditions or preferred approaches. But here, as elsewhere, there are limits to expertise that must be acknowledged. On the other hand, the term "convergence" does justice to the fact that most of the members in the groups were not didacticians.

The layout of the proposals made by the GTD's is in itself proof of the recognition that, to orientate a teaching cycle in a given discipline in a significant manner, it is not enough to make a list of all the changes planned. At the instigation of the CNP, the syllabus for each grade includes an introduction defining "guiding principles" and "general objectives". For every item in the syllabus itself, the "expected competences" are defined. And within this syllabus, the Physics and Chemistry groups have provided a menu of "back-up activities" (Physics), or "documentation or experimental activities" (Chemistry). And in addition to the official published "comments," accompanying texts were sent to Regional Educational Inspectors and to the organisers of teacher training programmes. These provide details on sensitive questions, and suggest approaches, sample tests and lists of materials. Publishers of textbooks were invited to discuss with the members of the Physics and Chemistry groups the

[18] I.e., "Quatrième," third year of secondary education.
[19] "Terminale": the last year of secondary education.

intentions behind their proposals, and by what means their recommendations could be implemented. It has, in short, been a very elaborate process.

In France, grade 8 has been, since 1993, the first grade in which Physics and Chemistry are studied as disciplines in their own right (*Bulletin Officiel*, 1992a). Only a few aspects can be approached here: objectives, the role of experimentation, conceptual progression, and how far common reasoning was taken into account in the decisions reached.

These points can be discussed only briefly here, but it is important that they be raised. Indeed, any attempt at orientating teaching means specifying and adjusting an entire set of parameters, not just one – not appreciating this means that without acknowledging it or even being aware of it, one is in all likelihood working on one parameter in such a way as to affect all the others.

DISTRIBUTED OBJECTIVES

Among the guiding principles formulated for the whole of secondary education are some intentions that had been stated previously: promoting rigour, scientific method, and curiosity, introducing a variety of techniques, anchoring learning in everyday experiences and modern technology, and giving a coherent representation of the universe, notably by establishing links between several disciplines. The essential role of experimental activities is reaffirmed. The only surprising thing is that there should be such an accumulation of intentions: can one really hope to realise them all? In fact, the official texts separate out these ambitions.

In the introduction to the grade 8 syllabus, emphasis is placed on curiosity, acquiring technical know-how, developing a taste for a rigorous analysis of phenomena, learning to differentiate what one can explain in part from what "makes no sense at all". The authors also speak of "drawing personal conclusions from experiments," of "rigorous reasoning based on a few simple rules," of the "permanent validity of physical laws," and of the "dawning confidence [of the pupils] in their own capacity to make predictions and to put them to the test." Thus, technical know-how and conceptual capacities are considered, the latter centring on the idea that things don't just happen at random: there are laws ("a few simple rules"). There is no getting away from these laws, and in return they make it possible to predict events, to "reason" from experiments, to spot phenomena that can be "explained"; they encourage pupils to think about a problem rather than saying the first thing that comes to mind. However, the limited validity of models is not explained at this level. Thus, some choices are indeed made, at least in the intentions that are expressed.

All of this is divided up a little differently in the two parts of the syllabus.

Optics is particularly organised around the idea of conceptual coherence. There are two "simple rules":

"To be seen, an object has to send light to the eye";

"Barring accidents (an obstacle, a change in medium), light travels in a straight line".

These are expressed in everyday speech. In terms of abstract logic, these two rules can serve to introduce all the points later approached in optics. The course is conducted through a process of conceptual chaining; each concept introduced is first a target, and then a support, for learning. Thus the pupils can, or so one hopes, be brought to see (directly, at first) and to localise real images formed with lenses, and even to accept the idea that a mask on the lens affects only the brightness of the image. In short, the number of phenomena/number of "rules" ratio is particularly high, and there is no break in the chain of conceptual construction.

The preamble to the official instructions mentions few expectations concerning technical know-how in optics, stating only that care must be taken in the manipulations (some are, in fact, quite tricky).

As regards electricity, there is a better balancing of conceptual and operational aspects (i.e., schematisation techniques and measuring procedures). The conceptual structure is less linear here, and the course deals with two separate units, discharge and quasistatic changes. In the second part, the systemic aspect of the circuit is developed (for example, the order of the elements of a series circuit does not matter, the voltage of the generator is divided between the components in series), and, as always, the permanence of physical laws is stressed: "Add a lamp to a series circuit, and the values of the quantities change, but the laws remain the same". The second part also proposes an operational introduction to two quantities, current and voltage – they can be measured in a given way, with a given device, in a given unit. "Doing an experiment," "measuring" and "identifying" are among the "expected competences."

In terms of objectives, then, one cannot say that a syllabus of this type is monolithic. The intentions of the designers are varied and some are given more weight than others. But globally, at this grade level, the conceptual ambitions are quite high. Not so much because of the accumulation of items, but because the epistemological aims are set high: the idea of physical law is established, and attention is paid to sensitive points.

We shall come back later to this "sensitivity", as regards the pupils' difficulties. As for the idea that there are laws that must be taken seriously, and not just random phenomena, it may not be the direct object of research in didactics, at least as it is generally expressed. But its value as a teaching objective is obvious, in view of the results obtained by researchers on common conceptions and reasoning, and underlies nearly all their proposals for pedagogical intervention. Indeed, what, if not a constant demand for coherence, can help modify learners' most deeply rooted views? It is clear that experimentation in itself is not enough.

THE ROLE OF EXPERIMENTATION

Experimentation in physics teaching is always held up as an indispensable step towards clear knowledge in the field. That students should adopt the "experimental approach" is something everyone agrees on... It is more difficult to settle on the role the proposed experiments should be given. What do the official texts say? Explicitly, very little. Among the guiding principles, one finds the idea that if students can manipulate devices themselves, they will be "more involved" and hence "more responsible for building their own knowledge", but this reference to constructivism is all that one finds. The introduction for grade 8 (see the stated intentions, above) suggests more precise intentions: to associate experimentation and reasoning, phenomena and a unified vision. But the rationale behind the published intentions is best seen in the correspondence between the organisation of the syllabus and the proposed back-up activities.

Let us limit ourselves to optics. In this field, manipulation is of crucial importance. The pupils do not simply watch experiments, or even discover what laws apply. Here, phenomena can be interpreted in the light of prior knowledge and a newly learned law, and students can repeatedly make predictions, then observe and discuss the experiments, with the aim of establishing coherence. The experiments are not conducted to illustrate laws, but to show that it would be wrong not to derive conclusions from them.

For example, after the introduction of diffusing screens, rectilinear propagation and shadows, it is proposed that a relation be established between two situations: in one, shadows are cast on a lighted screen, and in the other, an observer is looking through a small hole in the screen. If the eye is behind the darkest part (the shadow), the observer will not see the associated source of light. From the semi-darkness, he/she will see part of the source. And from the lighted zone, he/she will see it completely.

The reference to constructivism is echoed, here, by a conceptual activity which goes far beyond manipulation or even inductive contemplation.

CONCEPTUAL PROGRESSION AND CONTENT ANALYSIS

All in all, the proposed themes are not particularly novel:

- sources, diffusion and colour,
- the rectilinear propagation of light and shadows,
- the eye and perception,
- (real) images formed by converging lenses.

It is the association of two approaches that is new. The contents have been revised according to a new hierarchy of concepts leading from one to another. And the common forms of reasoning of pupils and teachers are taken into account.

The content analysis re-establishes the eye as a privileged instrument. Hence the importance of the experiment described earlier, on looking through the holes of a screen: the point is to associate light and vision, not just by assertions, but in such a way as to allow the students to make non-trivial predictions. Moreover, the eye is essential as an instrument to construct the position of a real image, without a screen, by direct viewing. Purely perceptual aspects (often called "optical illusions") associated to the retina-brain unit, outside of the domain of geometrical optics, are also broached.

There is one further development in the content analysis: a critical approach to "rays" visualised by using a material support. These thin beams, classically used to introduce the rectilinear propagation of light, are subjected, here, to explicit and deferred analysis after introductory work on diffusion and shadows. Thanks to these official texts, teaching may feature fewer "fountains that are luminous because light cannot get out," or "rays that are visualised" by sheets of paper that do not include the source (Kaminski, 1989).

That the pinhole camera is no longer used as an introductory device is due to analysis of this type. Though relatively simple to set up, the device is, in fact, conceptually very complex, and the end result hardly justifies its use (see appendix 2).

COMMON FORMS OF REASONING AND THE CHOICES MADE

The proposed conceptual progression is remarkably consistent with analyses of common reasoning. Indeed, investigations[20] have shown that the role of the eye in vision is very inadequately taken into account in the reasoning of teenagers or even adults, including

[20] See Tiberghien (1984a), Guesne (1984), Fawaz (1985), Fawaz and Viennot (1986), Goldberg and Mac Dermott (1987), Kaminski (1989, 1991) and Chauvet (1990), among others.

teachers or advanced Optics or Art students. Thus, there is a need for explicit and focused work on this subject.

Yet, although the plan is to work on integrating the eye into the chain of vision, this has not been developed outright in the conceptual progression. That difficulty *has* been taken into account and is treated at length, but later.

The beginning of the progression takes into consideration another aspect of common reasoning – this time it is not an obstacle, but an aid to understanding. When an area of a screen is "luminous," "lighted" or "bright," everyone is convinced that light is reaching it. This correct idea is used to introduce diffusion by an object. Instead of asserting that "since we can see this area as lighted, it is sending light into the eye," the teacher places a second screen near the lighted area, to make the diffused light manifest. To render such a demonstration more effective, the first diffusing screen used is a sheet of coloured paper, and it is illuminated with white light. The second screen, which is "normally" white (in white light) takes on the same colour. The aim is to convince the pupils that diffused light does exist. All kinds of work on the different sorts of screen then become accessible, with colour as a stimulating support. This is also a very pertinent way of introducing instruction on colour.

Common forms of reasoning are thus taken into account either as bases or as obstacles which are explicitly targeted in teaching. Finally, there are common forms of reasoning that, it is hoped, our teaching will not reinforce, such as the idea that an image travels as an entity and arrives on the screen even without a lens, or, when there is a lens, the idea that a hole will be punched in the image by a diaphragm on the lens.[21] The approach[22] and the situations proposed here should make such a view of things less prevalent among pupils.

To sum up, this is how common reasoning is taken into account in this part of the syllabus:

Main ideas introduced	Aspects of common reasoning that are taken into account	Role of these aspects in the progression
Sources, diffusion, colour.	Belief that if an area on a screen is "luminous", that means it is receiving light.	Back-up.
Rectilinear propagation, shadows.	Same as above.	Back-up.
For an object to be seen, it must send light into the eye.	Implicit denial of this necessity. Belief that light can be seen from the side.	Obstacle and target, not reinforced by an early use of materialised rays; treated with the support of areas of the screen illuminated to a greater or lesser degree.
Perceptual effects.	Underestimation of the role of the eye-brain unit in normal vision.	Obstacle and target, to support the treatment of the previous concept.
Real image in a converging	Reasoning: it is "as if" the	Non-reinforced obstacle,

[21] Regarding the "travelling image", see Fawaz and Viennot (1986), Feher and Rice (1987), Goldberg and Mac Dermott (1987), Kaminski (1989), and box 5, chapter 2.

[22] This grade 8 syllabus is very close in many respects to the proposals of Kaminski (1991).

Main ideas introduced	Aspects of common reasoning that are taken into account	Role of these aspects in the progression
lens: how to see it and construct its position. A small part of the lens is enough to form the image.	image "travelled" all in one piece.	treated with the support of prior knowledge on the role of the eye.

CONCLUSION

The 1993 text on the optics syllabus at grade 8 and the accompanying texts illustrate undeniable convergences between the designers' views and didacticians' conclusions. The results of research have been taken into account. The official texts adopt two approaches that are crucial to didactical research: the analysis of conceptual content and attentiveness to common forms of reasoning. The designers' intentions and the didactitians' considerations also concur on the role that should be ascribed to experimentation, and, more generally, on the idea of associating a strong conceptual aim with limited mathematical formalism, at least initially.

The next task will be to assess the actual results of the most innovative aspects of the official proposals. This demands considerable follow-up work, testing and progressive adjustments, to be carried out over some time, for no one is naive enough to believe that the ideal solution can found at the first attempt, no matter how thorough the preliminary research and reflection.

But, most importantly, it takes more than just a few texts to inform and convince teachers, especially when the texts deal with matters on which it may seem that there is nothing left to say.[23] Here we wish to point out, once again, the importance of in-depth training programmes for teachers, in which didactic arguments are developed. It is not enough simply to present them with a few directives.

[23] See Hirn (1995) and Couchouron, Viennot and Courdille (1996), and more recently: Hirn and Viennot (2000).

APPENDIX 2

EXCERPT FROM THE ACCOMPANYING DOCUMENT FOR THE FRENCH SYLLABUS AT GRADE 8, IMPLEMENTED IN 1993

(Physics GTD, 1992)

WHY THE PINHOLE CAMERA IS NO LONGER ON THE SYLLABUS

The pinhole camera is not on the syllabus. This might seem surprising, considering it is a popular and traditional device, and one that is easy to construct.

It was dropped because of the following drawbacks:

The new syllabus was devised for pupils to construct the notion of an optical image, linked to the notion of perfect point to point correspondence; in this concept, each object point corresponds to one single image point and vice versa (this, obviously, is never completely realised). Defined in this way, the notion differs from an image taken in the larger sense of a "representation of an object" (as regards the definitions of the term image, see P. Léna and A. Blanchard, 1990, chapter 3).

The precise (and "ideal") notion of an optical image implies a localised image. The fact that "all the rays stemming from point A and passing through the optical device converge at another point A'" makes this localisation necessary.

In this respect, the pinhole camera gives a representation of an object, not an optical image. Actually, the disadvantage of the pinhole camera is not that it is an imperfect device: they all are, since the notion of an optical image is precisely a "borderline" notion. Rather, it is the very principle of the pinhole camera that is in question, since the pinhole camera is closer to the notion of a shadow than to that of an optical image. What is often referred to as an "image" with this sort of device is not localised, and is not an optical image. In particular, the eye, when placed in the luminous beam beyond the back of the pinhole camera, would not distinguish the form of the source by looking in its direction, if the bottom of the pinhole camera were removed (this is not due to weak light). On the other hand, without the screen, a real optical image formed by a lens is perfectly visible under the same conditions.

The construction of the concept of an optical image can, as we pointed out earlier, be guided by pedagogical activities that establish coherent links between objectives: a target objective for one sequence – the rectilinear propagation of light, for example – can (one hopes) be used as a conceptual basis for the construction of the next idea – i.e., the role of the eye in vision, on which, in turn, rests the construction of the notion of the point by point object-image correspondence. The only logical place to insert the pinhole camera in this framework is in the study of rectilinear propagation and shadows. But as regards shadows, there are objects and experiments which are much simpler to interpret. Many experiments can be made that are both motivating and surprising. Adding a more complicated one would uselessly weigh down the syllabus. Indeed, the pinhole camera is a device that is easy to set up, but difficult to interpret; moreover, it is likely to reinforce erroneous ideas if it is not used with great care.

IT IS DIFFICULT TO INTERPRET:

Just how difficult it is to pass from the continuous to the discontinuous becomes apparent here. The difficulty is dealt with differently in the object-space and in the "image"-space. In the object-space, as was the case with the lens, the source is analysed as a set of points, but in the "image"-space, the lighted areas are superposed to form the representation described above. One might ignore this aspect (again, our objections are based on principle, not on the "imperfection" of the device). But the problem arises once more when the hole in the pinhole camera is widened.

IT IS LIKELY TO REINFORCE ERRONEOUS IDEAS:

How do pupils react to this sort of conceptual complexity? Surveys in various countries, including France, have shown that, after instruction with the pinhole camera, the great majority of pupils are not able to establish a contrast between the type of "image" obtained on the back of the pinhole camera and an optical image. They cannot draw a diagram to explain the formation of an "image" of an extended object by a "small" (but not a "pin-point") hole, much less predict what will happen with a wider hole.

The answers obtained do, however, prove the popularity of the idea of the "travelling image", i.e., forms of reasoning in which the image is pictured as moving as a whole; obstacles (notably masks on lenses) are imagined as removing bits of it as it passes (a coin placed on a thin lens would, according to this type of reasoning, make a "hole" in the real image of an object), and lenses are thought to invert the image. The pupils say, for instance, that "the image takes the shape of the hole" if it is a big hole, or that "it passes through the hole, turning around" if it is a small hole (to get through it better?), or that, in the absence of any optical device, the image of a source will fall on the screen "erect, because it is not hindered by any optical device."

IT MUST BE USED WITH GREAT CARE:

Of course, the pinhole camera is not a "definite pedagogical DON'T" for all that. Its use is sometimes justified: for example, it can help to make clear why spots of sunlight on the ground are always round, even though the spaces between the leaves are not. But it calls for careful analysis, to compare what takes place when a lens is placed over the hole and when it is removed, and this would take too long at grade 8-level. The pinhole camera therefore appears more useful as a supporting device for a synthesis of elementary optics than as an introductory device.

REFERENCES

Bachelard, G. 1938. *La formation de l'esprit scientifique*. Vrin, Paris.

Bachelard, G.1966. *Le rationalisme appliqué*, PUF Paris (1949.)

Bulletin Officiel du Ministère de l'Education Nationale 1992a, n°31, Classes de quatrième et quatrième technologique, pp 2086-2112.

Chauvet, F. 1990. *Lumière et vision vues par des étudiants d'arts appliqués*, Mémoire de Tutorat non publié (L.D.P.E.S.), D.E.A. de didactique, Université Paris 7.

Chauvet, F. 1993, Conception et premiers essais d'une séquence sur la couleur, *Bulletin de l'Union des Physiciens*, 750, pp 1-28.

Chauvet, F. 1994. *Construction d'une compréhension de la couleur intégrant sciences, techniques et perception: principes d'élaboration et évaluation d'une séquence d'enseignement*. Thèse. Université Paris 7.

Chauvet, F. 1996. Teaching Colour : Designing and Evaluation of a Sequence, *European Journal of Teacher Education*, vol 19, n°2, pp 119-134.

Couchouron, M., Viennot, L. and Courdille, J.M. 1996. Les habitudes des enseignants et les intentions didactiques des nouveaux programmes d'électricité de Quatrième, *Didaskalia,* n°8, pp 83-99.

Driver, R., Guesne, E. and Tiberghien, A. 1985. Some Features of Children's Ideas and their Implications for Teaching, in Driver, R., Guesne, E. et Tiberghien, A. (eds): *Children's Ideas in Science*. Open University Press, Milton Keynes, pp 193-201.

Fawaz, A. 1985. *Image optique et vision: étude exploratoire sur les difficultés des élèves de première au Liban*. Thèse de troisième cycle. Université Paris 7

Fawaz, A. and Viennot L. 1986. Image optique et vision, *Bulletin de l'Union des Physiciens*, 686, pp 1125-1146.

Feher, E., and Rice, K. 1987. A comparison of teacher-students conceptions in optics, *Proceedings of the Second International Seminar: Misconceptions and Educational Strategies in Science and Mathematics*, Cornell University, Vol II, pp 108-117.

Galili, Y. 1996. Students' Conceptual Change in Geometrical Optics, *International Journal of Science Education*, 18 (7), pp 847-868

Galili, Y. and Hazan, A. 2000. Learners' Knowledge in Optics, *International Journal of Science Education,* 22 (1), pp 57-88.

Goldberg, F.M. and Mac Dermott, L. 1987. An investigation of students' understanding of the real image formed by a converging lens or concave mirror, *American Journal of Physics*, 55, 2, pp 108-119.

Groupes Techniques Disciplinaires de Physique et de Chimie 1992. Avant-projets des programmes de physique et chimie, *Bulletin de l'Union des Physiciens* , 740, supplément pp1-52.

Groupe Technique Disciplinaire de Physique 1992. *Document d'accompagnement pour la classe de quatrième,* Ministère de l'Education Nationale et de la Culture.

Guesne, E., Tiberghien, A. and Delacôte, G. 1978. Méthodes et résultats concernant l'analyse des conceptions des éléves dans différents domaines de la physique. *Revue française de pédagogie*, 45, pp 25-32.

Guesne, E. 1984. Children's ideas about light / les conceptions des enfants sur la lumière, *New Trends in Physics Teaching*, Vol IV UNESCO, Paris, pp 179-192.

Hirn, C. 1995. Comment les enseignants de sciences physiques lisent-ils les intentions didactiques des nouveaux programmes d'optique de Quatrième? *Didaskalia*, 6, pp 39-54.

Hirn, C. and Viennot, L. 2000. Transformation of Didactic Intentions by Teachers: the Case of Geometrical Optics in Grade 8 in France, *International Journal of Science Education*, 22, 4, pp 357-384.

Kaminski, W. 1986. *Statut du schéma par rapport à la réalité physique, un exemple en optique*, Mémoire de tutorat, D.E.A. de didactique, Université Paris 7.
Kaminski, W. 1989. Conceptions des enfants et des autres sur la lumière, *Bulletin de l'Union des Physiciens* , 716, pp 973-996.
Kaminski, W. 1991. *Optique élémentaire en classe de quatrième: raisons et impact sur les maîtres d'une maquette d'enseignement*, Thèse (L.D.P.E.S.), Université Paris 7.
Lijnse, P., Licht, P., de Vos, W. and Waarlo, A.J. 1990. *Relating macroscopic phenomena to microscopic particles, a central problem to secondary education.*CD-B Press Utrecht.
Léna, P. and Blanchard, A. 1990. *Lumières. Une introduction aux phénomènes optiques.* Interéditions Paris.
Tiberghien, A. 1984a. Revue critique sur les recherches visant à élucider le sens de la notion de lumière chez les élèves de 10 à 16 ans, recherche en didactique de la physique : *les actes du premier atelier international, La Londe les Maures, 1983*, CNRS, Paris, pp 125-136.
Viennot, L. 1994. Recherche en didactique et nouveaux programmes d'enseignement: convergences. Exemple du programme de Physique de quatrième 1993 en France, *Didaskalia* 3, pp 119-128.

Chapter 3

The real world: intrinsic quantities

The help received from Edith Saltiel in the preparation of this chapter is gratefully acknowledged.

Is the fact that concepts are quasi "materialised" a great obstacle to physical analysis? Are the resulting gaps between common thought and orthodox theory the cause of misconceptions that go beyond optical images being slightly too solidified or a notion of colour that is too closely linked to painting?

The answer is yes, definitely. This is especially evident when we consider an idea that is essential in physics: that physical quantities do not exist in themselves. They are defined and they are measured, and this involves (at least) the following aspects: a frame of reference, a unit, and the evaluation of uncertainty. Let us limit ourselves to the first of these points.

1. THE ESSENTIAL: DEFINING A FRAME OF REFERENCE

In physics, it is necessary to define quantities that can characterise phenomena, for example, mass, position, velocity, and so on. So one must define what is called a frame of reference. An origin, three unit vectors and some clocks are required to ascribe to any event a place and a date that can be recorded in numbers. In this way, an event can be located by its position in space (x, y, z) and time (t) in a given frame of reference, and by the corresponding coordinates x', y', z', t' in another frame of reference.

If the relative velocities are "ordinary" (i.e., not of the same order of magnitude as the speed of light), one can assume that the time is universal. The time of the event is then identical in the different frames of reference.

Otherwise, one must apply Einstein's theory of relativity, which poses enormous problems to the intuition because it questions the idea of simultaneity.

Let us limit ourselves here to Galilean relativity, which is adequate for the mechanics of common objects.

To get an idea of what a (two-dimensional) frame of reference is, imagine a camera whose images are dated and have two axes that are always in the same place on the film. The events recorded will not be the same if two such cameras are set in different places, especially if one is moving relative to the other.

But can a frame of reference "move"? To answer that question, one would have to know what it is to be absolutely stationary: impossible! In order to prove that one is not moving, it is necessary to refer to another point... and what if it, too, were moving? There is, therefore, no absolute immobility, and no frame of reference is more absolute or more immobile than any other. Are they all the same, then? No. The great founders of modern science, Galileo, Newton, Huyghens, and all those who accompanied them in their discoveries, established that in certain frames of reference, one could successfully apply the simple laws of classical mechanics – in particular, the principle of inertia, according to which the velocity of a body on which no force is exerted is constant in its magnitude and direction (more precisely, this applies to one point: the centre of mass of the body). These frames of reference have since been termed Galilean. If one is known, it is possible to know them all: each one moves in a straight line relative to any other, with a constant velocity and without changing the direction of the axes. Such frames are very convenient, because Newton's laws apply to them easily. Others are often referred to as "accelerated" frames of reference, for the sake of brevity.

It is common practice to choose one frame – preferably Galilean – and to adhere to it. But it is sometimes necessary to compare the description of a phenomenon in one frame with its description in another.

Let us imagine two parachutists holding cameras, filming the same things while falling at different velocities. Or two travellers in two different trains moving at different velocities, fascinated by the same cow, and filming it. Each time, the films will be different. Nevertheless, certain physical quantities will appear identical and others different. For example, each traveller will find the same value for the acceleration of the cow, but not for its velocity: it is said that acceleration is the same in all Galilean frames of reference, which is not case with velocity. Transformation formulae allow one to convert from the velocity vector found in one frame to that found in another. And this also applies to other physical quantities: linear momentum, kinetic energy...

The main thing is that, to a physicist, all these descriptions are of equal merit. None describes reality better than another, and the laws of physics apply to each in the same way. One can only say that such or such a frame is more convenient, because it makes for simpler calculations, or gives a better idea of a situation.

2. QUESTIONS: FISHES, PARACHUTISTS AND MOVING WALKWAYS

For more detailed information, see Saltiel (1978); see also Saltiel and Malgrange (1979).

Questions and correct answers to first year university students with several basically similar situations; that is to say that once the phenomena are translated into symbols, the equations and the reasoning involved in finding the correct solution are the same. Each time, the students are asked if two observers agree on the value of certain quantities.

In Galilean relativity, and for two observers in rectilinear relative motion at constant velocity

- the interval of time between two events has the same value;
- the velocity of a moving object has a different value;
- the distance covered by this moving object over a given time interval also has a different value (the relationship between the distance covered d, time t, and velocity v must be the same in both frames: $d = vt$ for uniform motion);
- on the other hand, an object's geometrical dimensions are the same for both observers (who know how to take into account distances and angles of vision, like all good physicists).

And so, to go over this list of statements from the bottom up, the length of a jumping fish is the same for the two swimmers, but the length of its jump is not the same (to a swimmer borne along by a fast current, it could even seem to be falling in the same spot), the velocity of the jump is not the same for each swimmer either; but their stopwatches indicate that the same time has elapsed.

All of this can be transposed to the other two examples, concerning parachutists and moving walkways (for the correct answers in detail, see box 1).

2.1 Results and comments

Many students mistakenly assert that velocity is invariant, for example: "The velocity of the jump of the fish is the same for the two swimmers" (30%, n=46). Likewise, the distance covered is erroneously understood to be as invariant as an object's dimensions: half of the students say that the length of the jump of the fish does not depend on the observer.

Box 1

Questionnaire on fishes, parachutists and moving walkways

For each exercise, the observers have at their disposal the means to measure the objects considered with the precision they require, and in their frame of reference.

First read the description of the three situations below:

Exercise 1

Two swimmers are floating in a river; each of them is hanging on to a separate buoy. One of them is unwittingly being carried along at a constant velocity by a strong current.
A small fish is jumping in a direction that is more or less parallel to the current.

Exercise 2

Two parachutists are falling vertically, each one at a different, but constant, velocity. One of them drops his/her eyeglasses, the other one catches them.

Exercise 3

In an airport, a moving walkway along the corridor enables travellers to reach the gates more quickly. There are two men, A and B: A is standing still in the corridor; B is on the moving walkway, leaning against the handrail. Both are watching a third man, C, who is walking on the moving walkway.

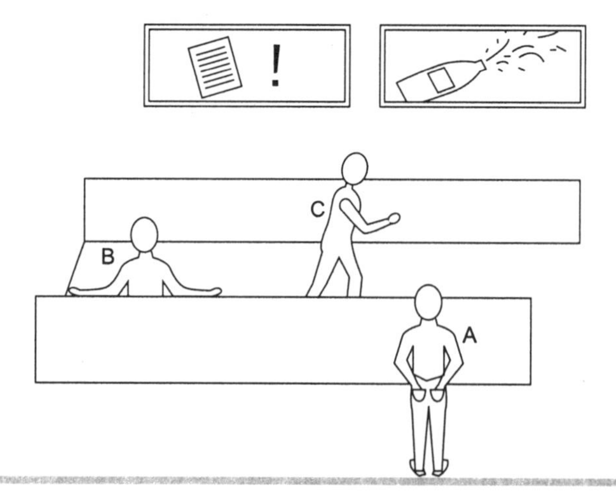

Now answer the following questions:

	Correct answer...
Exercise 1	
Are the following quantities the same for the two swimmers?	
Time the fish spends out of the water	Yes
Length of the fish	Yes
Length of its jump	No
Velocity of the fish at a given moment	No
Exercise 2	
Are the following quantities the same for the two parachutists?	
Duration of the eyeglasses' fall	yes
How far the eyeglasses fall	No
Size of a parachute	Yes
Velocity of the eyeglasses at a given moment	No
Exercise 3	
Are the following quantities the same for the two observers, A and B?	
Duration of one of C's steps	Yes
Width of a poster	Yes
Height of a poster	yes
Distance covered by C in one step	No
Velocity of a poster at a given moment	No
Velocity of C at a given moment	No

When the students do perceive certain aspects of Galilean relativity, for example the fact that the films mentioned in the questions are different, their first reaction is to prevaricate: "It is the visible motions that differ". These students, who are unwilling to accept Galilean relativity in its entirety, favour quantities that they consider to be "real" or "true." In their opinion, the velocity "supplied" by a motor – physicists would say, "defined relative to the support of the motor" – is a "true" velocity, and the corresponding trajectory and direction are also "true."

According to this logic, the other frames of reference only allow one to account for appearances. If, on one parachutist's film, the dropped eyeglasses fly back upwards, no one can be expected to take that motion seriously! The same goes for the poster pasted on the wall, which, seen from the moving walkway, appears to be moving backwards.

The idea that there is a true velocity for objects, "the" velocity of the object, with possible apparent variations depending on the observer, is one that researchers come across very often.

Furthermore, drag proves to be a decisive factor in the students' reasoning, though it is an aspect that physical theory is indifferent to. In the case of the moving walkway, a motor "moves" one frame of reference with respect to the other. This case is also the one in which the fact that velocity depends on the point of view of the observer is best taken into account. Seen from the fixed pavement, the efforts of two motors add up: the motor of the moving walkway and the muscles of the walking man. As a result, the velocities of the walking man, as estimated by the bystander on the fixed pavement and by the person leaning against the handrail of the moving walkway, are readily considered to be real, and the difference in their values is taken into account (only 10% error rate). The velocity of the fall of the parachutist's glasses is more often understood as being independent of the observer (36%): there is only one "motor," gravity, which makes it possible, seemingly, to define the "true" value of the velocity. The velocities of the objects as seen by the parachutists are thought to be merely "apparent" velocities.

The link between motion and motor therefore strongly determines the way velocity is perceived.[1]

3. WHEN DRAG DISAPPEARS...

There is a question that our ancestors found most intriguing. In Galileo's day, it was a matter of great debate whether a stone dropped from the mast of a moving ship would fall at the foot of the mast, or behind it. As early as the sixteenth century, Bruno[2] (1584) anticipated the answer:

> Whatever is on Earth moves with the Earth. The stone thrown toward the top of the mast will come back down no matter how the vessel moves.

But he needed the idea of an "intrinsic motion" – in other words, a cause internal to the moving object, to explain why the stone's motion in free fall is similar to that of the boat:

> ...The stone dropped from the hand of a person on board the ship, and consequently moving with the movement of the latter, possesses an "intrinsic motion."[3]

A furious intellectual battle lasted several decades. The following statement by Gassendi,[4] coming almost sixty years after Bruno's (in 1642),

[1] On this subject, see chapter 7 and Maury, Saltiel and Viennot (1977).

[2] Bruno, 1584, (1830) p 170, quoted by M.A.Tonnelat (1971, p 30): as Tonnelat stresses, Bruno was laying down the bases of the principle of relativity, which excludes "the estimation of the motion of a mechanical system through experiments realised from the system itself". See also Saltiel (1978).

[3] Bruno, De Motu, quoted by Koyré (1966), p 173.

[4] Gassendi, De Motu, quoted by Koyré (1966), p 316. See also Viennot (1979a), pp 113-130.

shows how far physicists had come by putting force back in its place, that is, in the motor:

> It would appear that the active force, which is the cause of the motion, is in the projecting agent itself, and not by any means in the projected object, which is purely passive. What there is in the projected object is in fact motion and, though it may be called force, impetus, etc. (terms we ourselves have used when, to make ourselves more easily understood, we kept to familiar language), it is never anything but motion itself.... Now, nothing prevents the received motion from continuing, should the motor detach itself or even die out. Because a motor is not required to transmit to the moving object any force besides motion; to produce motion, it is enough for it to provoke in the moving object a movement which can continue without the motor. Motion can do this, because such is the property of its nature, provided that the mobile remains, and that no contrary event affects it; it has the property of continuing without continued action from its cause.

This text is truly remarkable in that it not only expresses ideas which were to prove extremely fruitful, but also analyses the main obstacle to such an apprehension of motion: that it is not necessary to seek the "continuous action of the cause," because motion can continue without cause.

More than three hundred years later, we come across the same problem in the reactions of our students.[5]

For example, if a man on a moving walkway throws a ball "vertically" into the air, will it land in his hands? (See box 2 for an analysis of this situation). The many students who mistakenly answer "no" expect the ball to land behind the man since "while the ball is in the air, the walkway is moving forwards," or "when it is let go, it immediately loses its velocity." Once the physical link is broken, horizontal velocity disappears! (Saltiel, 1978). There must be a cause to explain the forward motion of the ball, and before it is released the cause is the moving walkway. One cause will do, no need to invent another: that would not be economical.[6] It does not even seem necessary to reformulate this cause in terms of force, as a recent study on friction[7] (box 3) has shown. But where does one go from there? If the walkway is no longer linked to the object, its action ceases, and along with it, the effect it is supposed to produce: the forward motion of the ball. This

[5] In spite of this, we do not support the thesis of a strict parallelism between ontogeny and phylogeny (the development of the individual and that of the species); see also Saltiel and Viennot (1985).

[6] See Gutierrez and Ogborn (1992) among others.

[7] Caldas (1994), Caldas and Saltiel (1995).

seems to be the train of thought leading to these common comments and erroneous predictions.

Now for what might be called "the optimistic physicist's interpretation". Having read the preceding lines, some readers will by now have begun to worry, thinking: "But the students are right, after all! The ball does in fact land behind the man, the air slows it down and it loses horizontal velocity while it is in the air. The walkway moves forwards more rapidly, etc." This is the subject of the following conversation; it was reconstituted from accounts of debates (Saltiel and Malgrange, 1979). The discussion concerns a film which some grade 12 pupils[8] have just seen, and which shows an actual ball being thrown in the air.

Here is a typical argument:

- Normally, the ball lands behind the man.

- Why?

- Because of the resistance of the air.

- What if the scene takes place in a train?

- Then the ball lands in his hands.

Box 2

Trajectory depends on the frame of reference

A ball is thrown into the air from a rolling walkway.
Two cameras, one (A) on the "fixed" corridor, the other (B) on the moving walkway, are filming the scene.

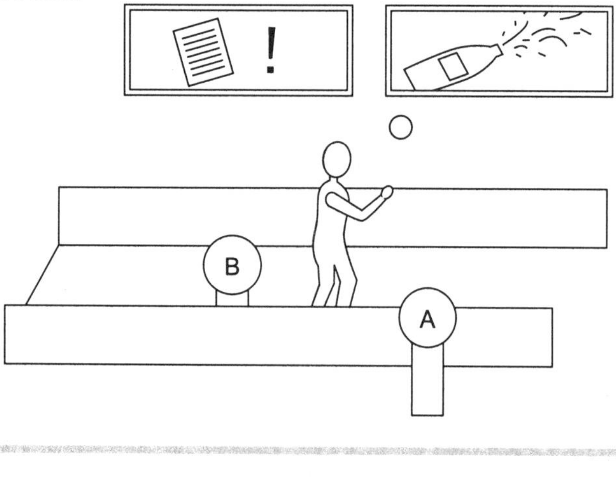

[8] I.e., Terminales, students in the final year of secondary education in France.

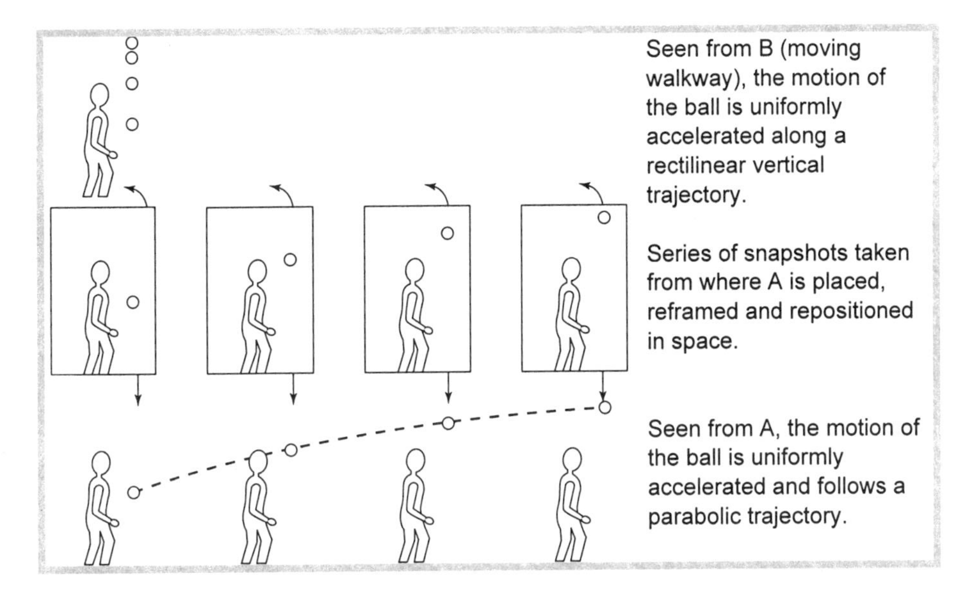

The answer is correct, says the optimistic physicist. But the investigator is not convinced:

- Why is that?

- Everything moves together, the air is dragged along.

- What if all the air in the train were removed?

- Then the ball would land behind the man again, there would be no air to drag it along any more.

Hence, the pupil correctly explains the fact that the ball lands in the man's hands by the forward motion of the ball (relative to the ground, its velocity being the same as the train's). But he/she needs to find a cause for this motion. Instead of admitting, like Gassendi, that motion "continues without continued action from its cause," the pupil attributes this to the drag caused by the air in the train. Although the answers are apparently correct, he/she is pursuing a line of reasoning whose bases prove to be those of common thought in the end.

Box 3

Drag due to friction: a situation in which it is not necessary to imagine a driving force

From Caldas (1994); see also Caldas and Saltiel (1995).

Outline of survey questionnaires

Asked to students or teachers in Portugal, Brazil and France (n=442), about the type of physical situation studied at school.

Some rectangular blocks are stacked on a table.

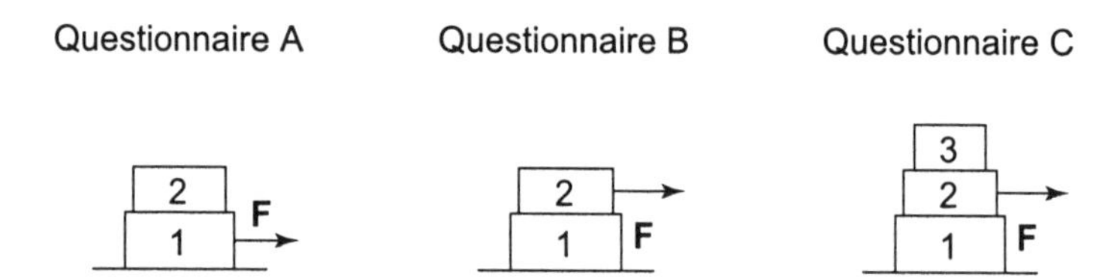

A constant force **F** is exerted on one of the blocks,, and the blocks then move with respect to one another. We assume that the only negligible coefficient of friction in the problem is the one corresponding to the contact between the table and the block resting directly upon it.

1. On the diagram, draw all the forces exerted on each of the blocks.
2. For each block, the participant is asked:
   - what forces are exerted on it?
   - what is the direction of motion of the block relative to the table?
   - is the friction force exerted upon the block opposed to its motion relative to the table, or not?

Common answers:

Approximately **two thirds** of the participants (regardless of nationality) **draw only one force at the interface**. Of these "single" forces, four fifths have a direction **opposite** to **F**.

Some comments provide more information on the particular status that students attribute to the friction force, for example: "The friction force is always opposed to the motion of the block", or "The force is a resistance" [a].

In the following representations of forces, no force is drawn for the blocks coloured in grey in the direction of the force **F**, although the answers state that the block will move in the direction of that force.

How do you explain this motion?

For these students,

the block that experiences the external force **F "takes the other one along with it", or "drags" it, through the effect of "friction", of "a tendency to follow," of "adherence," or of "friction that acts as a transmission belt."**
The **cause of the motion** is, at least in part, transmitted to the block that is dragged, **without the explicit intervention of a force** acting upon the latter in the direction of the supposed motion.

a. Caldas has also shown that the tendency to imagine that an object generates an opposition to "the motion of the block causing motion" is greatest when this object is beneath, not above, the block causing motion.

4. CONSIDERING NON INTRINSIC QUANTITIES: A TEACHING GOAL

The study of changing frames of reference confirms a common tendency to consider quantities as intrinsic characteristics of objects. Though justifiable for the geometrical dimensions of objects, this point of view leads to numerous errors in the case of velocity, distance moved, trajectory and direction of motion, which depend crucially on the frame of reference chosen for the description.

A bias towards the idea that a quantity is an intrinsic property of the object is apparent even when an answer is, at first sight, correct. Thus, when velocity ceases to be considered as intrinsic, the students generally attribute this to "drag," that is, the action of a motor on a support, two equally material elements of analysis. But when drag is seen as causing the motion of the object that is dragged, the idea inevitably arises that the corresponding velocity drops as soon as the physical link disappears ("The object immediately loses its velocity when it leaves the moving walkway").

All these difficulties are manifest when dealing with the "simplest" cases of changing reference frames: those involving Galilean relativity. There is no need, then, to bring up "rotating" frames of reference, let alone Einstein's special relativity, in order to measure the discrepancies between common sense and physical theory.

To state once again our position on this subject: we believe that, to be effective, instruction depends on pedagogical activities which make both teachers and pupils (or university students) more aware of difficulties; on a sustained vigilance regarding statements and formulations, especially those contained in textbooks, and above all, on clearly defined conceptual objectives for each sequence.

On the subject of reference frames, as for the preceding theme, the survey questions and the analyses provided, each dealing with a "sensitive problem" of understanding, can contribute usefully to an effort of this kind.[9]

But for debates to be fruitful, they must lead the student to search for a coherence greater than his or her own ideas offer. In kinematics, there must be some compensation for renouncing the seemingly determining role of drag: most often, it will be coherence rather than predictive correctness. The predictions are, nonetheless, generally not so faulty: common reasoning interprets facts in its own way, as far as possible, if necessary patching up favourite explanations with facts learned through experience. The example of the dialogue quoted above, on why a ball might land behind the person who threw it in the air, shows clearly enough that it is not such a simple matter to distinguish between a patched-up job and authentic Galilean reasoning.

Once an objective has been defined and adopted, what methods can be recommended to achieve it? The answer may come as a disappointment: the same methods as before, mostly. For, as regards this fundamental chapter of elementary mechanics, the content matter is clear, as are the pedagogical aids. For example, box 2 presents an analysis showing the points of view of two observers, which can be connected by a series of snapshots. This is very useful, and can be reworked with all kinds of more elaborate audiovisual materials: a CD-ROM would be ideal. Such a proposal will not come as a surprise to any teacher. But in fact, only the rigour of one's comments, the characterisation and the confrontation of habitual obstacles, and the stress laid on the fact that the same laws are valid in different frames of reference, will make these pedagogical aids effective. In other words, different frames of reference do not need to be explained by "new methods" but rather need to be taken seriously.

This does not mean that one should focus disproportionately on calculations. It is more a question of illustrating and respecting the corresponding concepts. If one has decided to introduce the non-intrinsic nature of velocity, for example, it is better to avoid the usual assertions on "absolute velocity" as opposed to "relative velocity." This can entail doing fewer calculations, but more constructions like the ones in box 2, particularly as regards two-dimensional motion, where one has to break with the "rigidity" of trajectories. And if one has only introduced the idea that the same object can have two different velocities, neither of which is any more or any less "fictitious" than the other, and that both velocities can be used in calculations based on the same theory, provided the frame of reference is clear, then that will be a great step forward in understanding what is essential in physics.

[9] See Saltiel and Viennot (1983).

But when gauging common thought, one finds evidence that the study of frames of reference is not the only instance in which reasoning in terms of quasi-material objects interferes with the consistent application of physical laws.

REFERENCES

Bruno, G. 1584. La Cena de le Ceneri III, 5 *Opere Italiane*, (Ed. Wagner, 1830).

Caldas, E., 1994. *Le frottement solide sec: le frottement de glissement et de non glissement. Etude des difficultés des étudiants et analyse de manuels.* Thèse. Université Paris 7.

Caldas, E. and Saltiel, E. 1995. Le frottement cinétique: analyse des raisonnements des étudiants. *Didaskalia*, 6, pp 55-71.

Gutierrez, R. and Ogborn, J. 1992. A causal framework for analysing alternative conceptions, *International Journal of Science Education.* 14 (2), pp 201-220.

Koyré, A. 1966. *Etudes Galiléennes* (p 136). Hermann Paris.

Maury, L., Saltiel, E. and Viennot, L. 1977. Etude de la notion de mouvement chez l'enfant à partir des changements de repère, *Revue Française de Pédagogie*, 40, pp 15-29

Saltiel, E. 1978. *Concepts cinématiques et raisonnement naturels: étude de la compréhension des changements de référentiels galiléens par les étudiants en sciences.* Thèse d'état.Université Paris 7.

Saltiel, E. and Malgrange, J.L. 1979. Les raisonnements naturels en cinématique élémentaire. *Bulletin de l'Union des Physiciens*, 616, pp 1325-1355.

Saltiel, E. and Viennot, L. 1983. *Questionnaires pour comprendre*, Université Paris 7 (diffusion L.D.P.E.S.).

Saltiel, E. and Viennot, L. 1985. What do we learn from similarities between historical ideas and the spontaneous reasoning of students?" *The many faces of teaching and learning mechanics.* In Lijnse, P. ed. GIREP/SVO/UNESCO, pp 199-214

Tonnelat, M.A. 1971. *Histoire du principe de relativité.* Flammarion, Paris.

Viennot, L. 1979a. *Le raisonnement spontané en dynamique élémentaire*, Hermann, Paris.

Viennot, L. 1979b. Spontaneous Reasoning in Elementary Dynamics, *European Journal of Science Education*, 2, pp 206-221.

Chapter 4

The essential: laws for quantities "at time t"

In association with Laurence Maurines.

1. INTRODUCTION

Common reasoning often involves stories about objects and the simple properties of objects; physical theory, on the other hand, has had to devise entire series of concepts – notably, physical quantities. Despite a constant concern for economy, it has proved impossible to reduce beyond a certain point the number of physical quantities that it is useful to define. Thus, in classical mechanics, one can hardly do without the quantities velocity (**v**), acceleration (**a**), mass (m), force (**F**), linear momentum (m**v**), energy (E), and work (W). A physical analysis of any dynamic situation therefore involves clearly differentiated concepts.

Some of these quantities concern particles (for example, velocity) or objects (for example, mass and kinetic energy). Others which involve an interaction require that we specify the two things involved in the interaction: Force of... on.... In this way, we can see that force is not a characteristic of an object, but the means of describing an interaction between two objects. This point will prove crucial in the difficulties commonly observed.

Nor should we neglect the role of time in the relationships between quantities established by theory. For the most part, they concern values at the same instant.[1] Indeed, this is so often the case that no one thinks of specifying it.

The two fundamental laws of Newtonian dynamics can be written as follows:

F=m**a** for a particle, **F** being the sum of the forces exerted on the particle (see box 1 on force, velocity and acceleration).

$\mathbf{F}_{(1)\text{ on }(2)}=-\mathbf{F}_{(2)\text{ on }(1)}$ for the reciprocal actions of two particles (1), (2) on one another.

But one could add, for every force term, the time variable, for instance: **F** (t)=m**a** (same t). The same applies to the law of reciprocal actions.

Here, then, are fundamental points of Newtonian mechanics. How are they understood?

2. ANALYSING THE MOTION OF MATERIAL OBJECTS: USUAL WAYS OF REASONING

From Viennot (1979).

A key point in the summary provided above is that Newton's relationship associates force and acceleration, in other words, force and the rate of change of the velocity vector (box 1). It is on this point that common reasoning diverges most markedly from taught physics.

2.1 Associating force and velocity

Box 2 sums up one of very first sets of results obtained in research on "spontaneous" reasoning.[2]

The situations shown involve forces which depend only on the relative position of one body with respect to the others. The question asks about the forces acting at a precise point in time on identical bodies placed in identical relative positions at the same point in time. In these situations, the forces depend solely on the positions of the bodies. They are therefore equal for the two systems considered, as is the acceleration of the moving objects.

[1] When, however, one applies a conservation relation, on one side of the "equals" sign is the value of the conserved quantity at a given time t_1, and on the other, the value at a subsequent time t_2. The difficulties linked to the need for a double localisation in time have long been recognised. Less attention is generally given to the difficulties arising in cases where the terms all concern a single instant.

[2] See also Viennot (1989a) and Mc Dermott (1984).

Box 1

Force, velocity, and acceleration: what are the relationships?

Newton's relationship associates the force (**F**) exerted on a particle of mass m with the acceleration **a** of that particle: **F**(t)=m**a**(t) (t: time)

The vector **a** is not directly accessible to the intuition, unlike velocity.

By definition, **a** is the rate of variation of the velocity vector **V** with time:

$$\mathbf{a}(t) = \text{limit of } \frac{\mathbf{V}(t+\Delta t)-\mathbf{V}(t)}{\Delta t} \text{ as } \Delta t \to 0.$$

It is possible to get an idea of this by examining, over a short time interval Δt, the change Δ**V**=**V**(t+Δt)-**V**(t) of the velocity vector. For example:

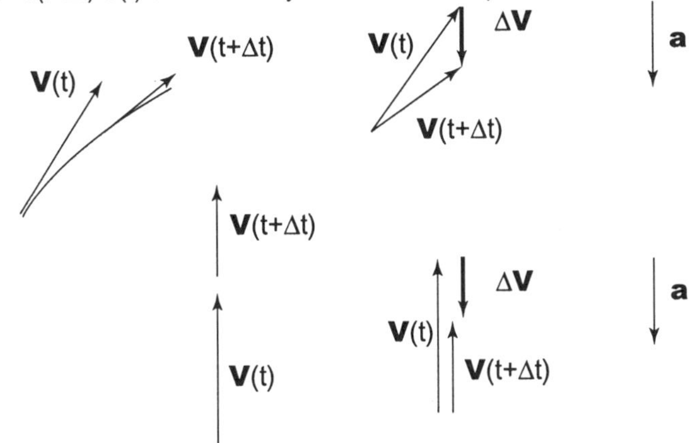

These two examples show that two moving objects with very different velocities and trajectories can have the same acceleration, and that the same force can therefore be acting on them (assuming that they have the same mass).

But in the situations presented, the velocities and, more generally, the motions, differ. This results in a surprisingly high frequency of answers like: "The forces are different because the motions are," or, "The force is zero because the velocity is zero." These students establish a relationship between force and motion, or, more specifically, between force and velocity.

Other comments also deserve our attention. They reveal:

- a delocalisation of physical quantities in space and time: "At the top of the trajectory, gravity and the force of the movement of the thrower are at work" (although the thrower is no longer accomplishing any action when the object reaches peak height);
- an attribution of force to the object: "The force of the mass upwards", "the mass has a force" (whereas in correct physics, mass is subjected to a force, exerts another force, but does not have a force of its own). This encourages the idea of a storage capacity, and of a temporal gap between cause and effect: "A supply of upward force", "the force the fellow gave it" (whereas only energy or linear momentum can be stored);

Box 2
Two questions about the same difficulty

Summary of the questions

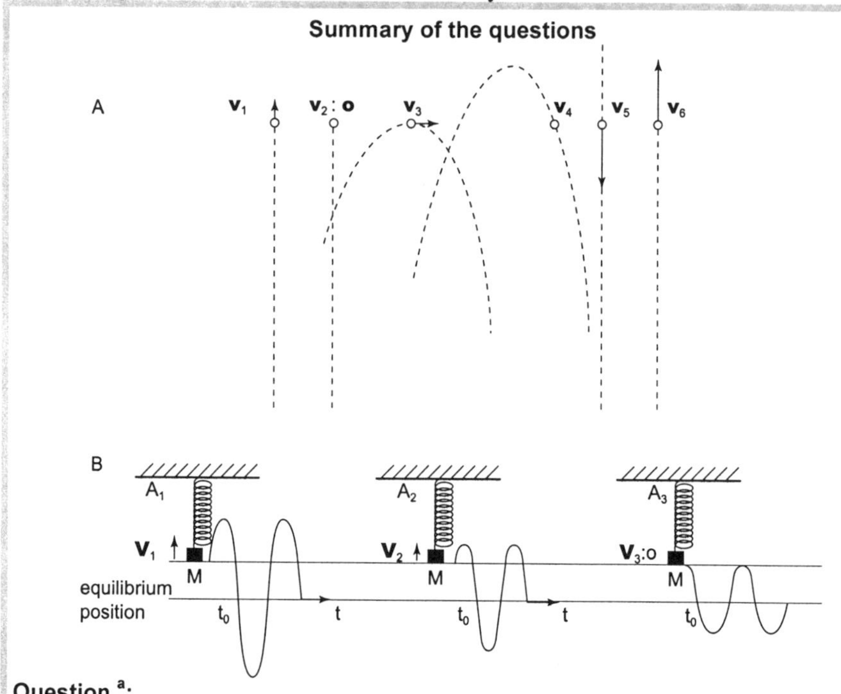

Question [a]**:**

A = a set of identical juggler's balls, all identical and at the same height at time t_0

B = identical oscillating masses, attached to identical springs of the same length at time t_0

In the situations shown above, are the forces acting on the different moving objects at time t_0 (the balls in A, the oscillating masses in B) equal, or not (assuming that air resistance is negligible)?

Explain your answer.

Correct answer

In both cases, the forces acting on the moving objects depend solely on their height. They are therefore **identical** at the **time considered**.

Rates of response

Number of students responding	Situation	Student's year of study	Answer: "the forces are not equal"
29		Last year of secondary school	55%
36	A	First year university	42%
226		First year university (Belgium)	54%
20		First year university	30%
95		Second year university	40%
49	B	Third year university	55%
14		Last year of secondary school (GB)	36%
14		First year university (GB)	43%
226		First year university (Belgium)	49%

Typical comments

"The forces are different because the motions are."

"The force is zero since the velocity is zero."

"The upward force of the mass..."

a. Questions A and B are presented together here, but were presented to the students separately.

- an indiscriminate use of the terms "force," "impetus," "velocity," "energy," "inertia," etc. (sometimes, diagrams are added in which a "force" and a "velocity" are added vectorially, in defiance of what physicists call "homogeneity");
- a certain vagueness about the ideas introduced: it is not clear whether these are represented by a (scalar) number or by a vector: "... a supply ... upwards" (force and velocity are vectors, but kinetic energy is a scalar, which has no "direction").

These characteristics, often found in situations involving "force-velocity", appear to indicate a view of physical phenomena that will require more to change it than the simple substitution of one relation (force-velocity) for another (force-acceleration).

2.2 Selective occurrence of some types of answer (Pendulum questionnaire)

Our previous hypothesis is supported by the fact that the frequency with which reasoning of the "force-velocity" type appears depends largely on certain theoretically irrelevant aspects of the situation proposed. These error-inducing factors are to be found in the questions cited above:

- The motion is "salient", i.e., it is presented in a form easily accessible to the imagination;
- It seems "incompatible" with the forces of interaction at work in the given situation: the known force does not have the same direction as the motion; or one of the two quantities, force and velocity, is zero, but not the other.

In other questions, where motion is presented in analytical form at the outset, or where only the forces are given, there are far fewer errors of this type.

An additional question (box 3) illustrates precisely the selective application of erroneous aspects of common reasoning. It presents four different kinematic situations for the same device: a simple pendulum, i.e., a small mass at the end of a string, the other end of which is attached. The forces acting on the mass in these four situations are considered. In all four cases, only the weight and the tension[3] of the string could possibly act on this mass.

These four situations do not result in the same proportion of "force-velocity" errors. Cases of "invented" force, aligned with velocity, become more frequent when real forces of interaction are less often mentioned (see columns 2 and 3 of the table in box 3). It would appear that the forces of interaction that one knows are more or less acceptable depending on the kinematic situation. The most perturbed case is when the mass is at the top of a complete circular trajectory, and experiences only downward forces: the tension of the string and the weight of the mass. Then, to a much greater extent than for the other three situations, the students feel a centrifugal force is necessary[4] to "balance" the weight. If the tension of the string is shown in the wrong direction, the desired radial balance can be obtained: "Zero radial velocity, therefore zero radial force."

[3] If the velocity is not zero, there is a centripetal acceleration and therefore also a centripetal component of force: the tension of the string adapts itself accordingly.

[4] These "centrifugal forces" cannot be interpreted by changing the frame of reference, as their occurrence is selective. Moreover, no change in frames of reference has ever "reversed" the tension of a string. The interpretation of the "optimistic physicist," who gives students credit for a correct justification (see box 7 below), does not hold.

Box 3
A more or less perturbed pendulum

Summary of the question

Draw a diagram of the forces acting on the mass in each of the following situations (column 1):

Situation	Kinematic data	Correct resultant	Common diagrams
P_1	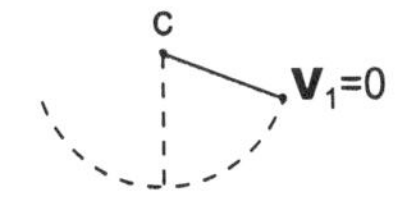 C $\mathbf{V}_1=0$	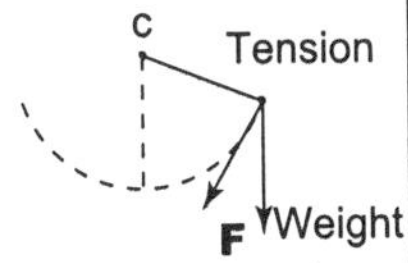 C Tension **F** Weight	 **F**=0
P_2	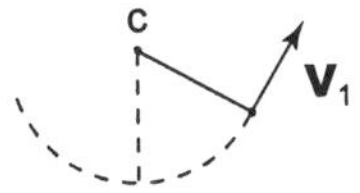 C $\mathbf{V}_1$	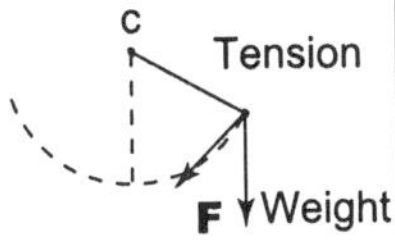 C Tension **F** Weight	 **F**
P_3	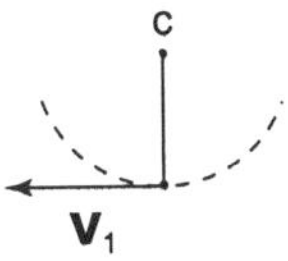 C $\mathbf{V}_1$	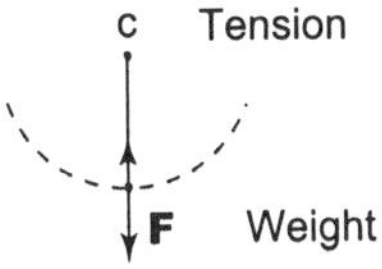 C Tension **F** Weight	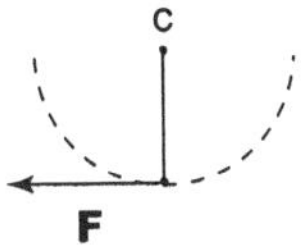 C **F**
P_4	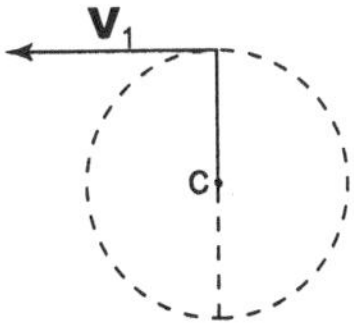 $\mathbf{V}_1$ C	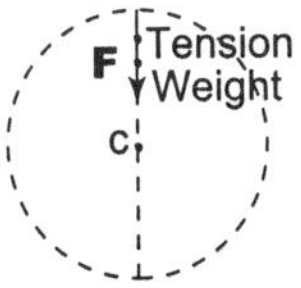 **F** Tension Weight C	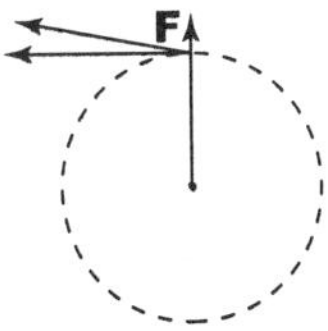 **F**

Outline of correct answers

- The tangential component of the resulting force is the projection of the weight of the moving object on the tangent to the motion. It is evaluated independently from the velocity of the moving object: its value depends on the position of the object.
- The normal component of the resulting force is the vectorial sum of the radial component of the weight and tension of the string. It can only be evaluated here from the motion that it contributes to determine. It is centripetal; its modulus is $\frac{mv^2}{l}$ (m=mass of the moving object; l=length of the string; v=speed).

Results

N=60	"F=V"	Tension and weight drawn	Tension reversed	Centrifugal force
P1	12%	78%	0%	0%
P3	26%	70%	0%	10%
P2	28%	58%	0%	2%
P4	37%	25%	12%	18%

Forces aligned with velocity (**F** - **V**) and "real" forces (tension and weight) in the students' responses.

- The less often interaction forces are mentioned (column 3), the more frequent are answers which "invent" forces aligned with velocity "**F** - **V**" (column 2).
- In the most perturbed situation (P4), references to an outward ("centrifugal" or "tension") force are most frequent.

Conversely, the case least favourable to error is when the mass is at its height, in maximum simple oscillation. True, the velocity is zero while the resulting force is not. Yet this force will soon have an effect: a slight gap in time between cause and effect is not a problem for common reasoning.

3. AN INTERPRETATION OF COMMON WAYS OF REASONING IN DYNAMICS

How can all these intuitions be combined to form a coherent whole? Here is one possible interpretation, based on simple ideas:

First, in the absence of motion, no questions arise.[5] But if there is an obvious motion, it has to be accounted for (see also Andersson, 1986). Sometimes, the idea that an object adheres to, or is part of, another moving object seems sufficient.[6] In some cases, drag is mentioned (see chapter 3) but it is not deemed necessary to identify the driving force acting on the object. It often happens, too, that a cause is found, such as "this pulls it... and it comes," "this pushes it... and it goes," "the weight brings it down... and it comes down" – then everything is all right.[7] But if "it goes up" when weight

[5] Séré (1982, 1985) has shown that this is especially true for gases: at the start of secondary education, teenagers cannot easily imagine the action of a gas on a wall.

[6] Gutierrez and Ogborn, 1992; see also the study on friction by Caldas, 1994.

[7] A variant of this type of analysis consists in interpreting by means of a single force of interaction what the physicists associate with the vector sum of two forces in opposite directions. The "suction force" so often alluded to in sailing manuals ought to make one wonder how molecules colliding on a wall, on a sail, or on the wing of a plane could ever cause anything but "push." To understand this so-called "suction," a difference between

is the only force acting, then there is panic. The same thing occurs in the case of the pendulum which "stays up in the air" even though two known forces are pulling the mass downwards.

What can continue to make things move, or make them "stay up in the air" in the absence of known forces? Their dynamism. This term might be used for the single notion to which the students haphazardly apply the more or less scientific terms enumerated above: "force," "impetus", "velocity", "energy", "inertia"... What matters is not so much the precise elements involved in the notion as its explanatory function. This "dynamism" is also attributed to the object: "The force of...."

Let us now examine the temporal aspect, which is very correctly connected with the idea of cause.

Cause ordinarily means the reason why things happen. If causes were persons, one might say they were responsible for events. In the definition of "physical causality," there is another element to be considered: cause precedes effect in time. In this sense, one cannot say that the fundamental relationship of dynamics, **F**=m**a**, presents force as the cause of acceleration, as it concerns quantities measured at the same time. A law associating simultaneous quantities is, to put it briefly... a law, not a story where causes precede effects.

But to speak of "the force of the thrower's movement, which acts at the top of the trajectory," corresponds to another type of logic, involving an initial cause. The idea of storage allows a previous cause to remain the reason for motion, or for the absence of falling, when one is needed: there is a delocalisation in time of the implicated quantities, for the purposes of the – causal – common explanation.

This type of temporal gap between quantities that should, in theory, "be taken at the same time" is very common in students' reasoning. In appendix 1, this tendency and its consequences are discussed further, in relation to a slightly more complex situation (concerning propulsion by a spring). Sometimes, too, the history of the motion manifests itself inappropriately through the trajectory, as though the circular or rectilinear nature of the preceding movement had been stored in the moving object (appendix 2).

The history of ideas contains traces of such conceptions. The pre-Galilean theories of impetus mentioned in chapter 3 are, by far, the most similar to these conceptions (Viennot, 1979). In most versions, a supply of "impetus" is stored within the object, which is thereby endowed with a greater or lesser "capacity for impetus," and this explains otherwise inexplicable movements. These coincidences are not raised here to support the simplistic thesis of a strict parallelism between the respective

---

two pushes should be envisaged. But one would then have to be willing to take two causal agents into account. (See the following chapter on the difficulties this entails.)

developments of the individual (ontogeny) and of the species (phylogeny). But they do throw some light on the strength of the trends of reasoning analysed in this book (Saltiel and Viennot, 1985).

The reductionist interpretation of physical phenomena, centred on representations of objects and the characteristics that one attributes to them –in this case, a dynamic supply of force which explains motion – is a familiar aspect of common reasoning. This may well be due, to a great extent, to our basic anthropomorphism.

4. COHERENCE AND RANGE OF COMMON WAYS OF REASONING IN DYNAMICS

Let us now examine how these aspects of natural reasoning extend to areas of physics which, from the point of view of accepted theory, have nothing in common with those dealt with so far. Common thought, however, establishes a link.

Such is the case with the propagation of mechanical waves, an issue introduced here through two examples: the propagation of a "bump" on a rope, and the propagation of sound.

4.1 Bumps on ropes

From Maurines (1986); see also Maurines and Saltiel (1988a).

Let us follow the course on a stretched rope of a bump caused by flicking it once with one's hand. The phenomenon is startling: the shape is maintained throughout its course; fragments of thread inform one another in turn, so as to reproduce the initial curve time after time. Physics teaches us that the speed of propagation of the signal depends solely on the rope and on its tension. But more important is the idea that what is moving is not a material object, made up of the same particles from start to finish.

How does common reasoning deal with this?

If students are asked if a bump moving along a rope can be made to move faster (box 4), many reply with the following type of comment:

"You have to shake your hand harder, the bump will have more force, it will move faster." A sketch reproduced in box 4 confirms that the force in question is in the bump, and directed forwards.

Box 4

Bumps on ropes

From Maurines (1986); see also Maurines and Saltiel (1988 a)

The force of the signal and its speed

Question

A red thread is tied to a rope laid on a horizontal table (at point R). A man takes hold of the rope at one end (0).

0 R

The man moves his hand. The following shape can be observed at time t:

0 R

Is it possible for the man to move his hand in such a way that the shape will reach the red thread before time t?
Justify your answer.

Outline of a correct answer

The speed of propagation depends solely on the medium (in this case, the rope: its mass, its tension). So any signal that does not modify these properties propagates at the same speed and, therefore, takes the same time to reach R.

Common answers

"Yes, he can move his hand to make the signal reach R sooner."
Students in grade 11 (science section)[8] (in France): 60%
First and third year university students, after instruction (in France): 75%

Students' comments

"The bump will move faster and faster if the hand movement is speeded up."
"The speed depends on the force with which the man moves his hand".
"If the magnitude of the force that is propagated is stronger, then the bump will move faster."

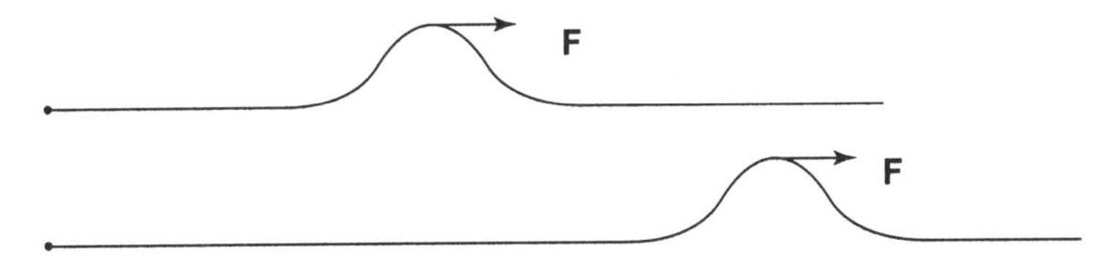

[8] Première scientifique.

A dwindling supply of force

For 68% of the pupils before instruction, and 55% of the university students after instruction, if the amplitude of a signal decreases with time, then its speed also decreases with time, for the following reasons:

"The height gets less because the hand movement slows down."

"If the bump disappears, that is because the force that created it has disappeared. At the same time, the speed decreases."

"The force produced by the hand runs out."

Again, considering two bumps travelling on the same rope, many students predict that the bigger one will overtake the smaller because "it has more force". Other remarks on bumps and ropes sound oddly familiar:

"The bump flattens, it has less and less force, so it moves more and more slowly"; and sometimes a student adds: "... because of friction."

Reading these statements, we can really see the "common dynamics" of solids at work. Here, an object first had to be invented: the bump. Then a cause had to be found for a bizarre, unexplained motion – the movement of the bump along the rope. It is the force stored within the bump, which wears out little by little. This force is "the force of the initial movement." Past cause and present reason are fused, a dynamic supply of force being attributed to the object: a surreptitious temporal delocalisation appears in the analysis.

4.2 Fast and slow sounds

From Maurines (1993)

The author of the preceding investigation completed it with a study on sound.[9] In view of the first results, it is easy to predict those of the second study.

Sound is another signal which moves with a speed that depends only on the state of the medium that it travels through. This medium may be a gas, a liquid or a solid, with various characteristics (density, temperature...), but certainly not a vacuum.

Yet, if sound is thought of as being an object, why shouldn't it propagate in a vacuum? That is, in fact, what a lot of students think: "In steel, sound does not propagate, whereas in a vacuum, there is nothing to impede its propagation"; or: "If there is an explosion on the moon, an astronaut in orbit will be able to hear it" – although there is no air on the moon. We should acknowledge, of course, that the vocabulary used to describe

[9] For more detail, see chapter 8.

communications between space shuttles, space laboratories, and the Earth can easily lead to confusion: they can be "listened to," or "heard".

In view of the logic illustrated twice so far, it comes as no surprise that the students should have decided that the louder one shouts, the faster the sound will travel: "If Pierre shouts louder than Jean, the sound will have more force, it will travel faster". Once again, it is to the initial cause – the source – that the subsequent property of the signal is attributed: it travels faster (than one sent off with less energy). The future tense, so often used in these explanations, is not merely a stylistic device. It stresses the idea of causality, in the physical sense of the term: cause precedes effect.

Pedagogically speaking, some definite conclusions can be reached.[10] When teaching the propagation of waves, it is not enough to repeat the refrain (which students easily learn by rote): "For a given medium, the speed of the signal is a constant." One must at least point out that this is surprising, even shocking: "The speed with which the bump moves does not depend on the initial motion"; the same goes for sound. The first statement implies the second, some readers will say. Logically, perhaps, but this is not the students' reasoning: the first statement is a refrain learnt at school, whereas the second one addresses – and, what is more, curbs – a natural tendency of thought (see also chapter 8). It gives meaning to the refrain learnt at school.

The predictive power of our analysis as a whole is confirmed here. The answers to these problems on propagation observed are comparable to those obtained on the dynamics of solids whenever the problem requires students to explain a very obvious motion without any obvious "cause." The initial cause is then seen as determining: it is stored in the object to allow it to catch up with its effect in time, through a temporal delocalisation that is never very explicit.

The deviations from accepted theory are not, however, the same in the analysis of signals as in the dynamics of ordinary objects. The solutions arrived at through natural reasoning are even more incorrect, so to speak, for situations involving propagation, where the motion predicted is wrong (the wave speed of the signal does not depend on the source), than for situations involving the motion of solids, where only the interpretation of the motion is missing.

But the rationale behind all the comments is surprisingly similar, whether the topic is a stone, a bump, or a sound. Though no school ever consciously teaches anything of this sort, these forms of reasoning are extremely enduring!

Returning to the dynamics of ordinary objects, we will now explore some further repercussions of these typical aspects of natural reasoning. This time,

[10] See also Maurines and Saltiel (1988a).

it is not the fundamental law of dynamics which is violated, but that of reciprocal actions.

4.3 Newton's third law: some problems

Attributing force to an object and, in some cases, creating temporal discrepancies, make it much more difficult to understand interactions between two objects –especially in a situation where both are moving. Newton's law stipulates that their reciprocal actions are of equal magnitude: the driver whose car has broken down and who is pushing it with all his might towards the garage experiences a force that is the exact opposite of the force he is exerting. Nor is the wrestler who gets the better of his opponent and pins him down an exception to this rule: he never exerts on his rival a force larger than the force that the latter exerts on him. A sketch specifying how each force is applied (box 5) shows how one can reconcile this surprising equalising law with the fundamental relationship of dynamics, **F**=m**a**, according to which the sum of the forces acting on an object must be greater than zero for the object to accelerate.

It shows the role that the ground plays.

But if the situation is seen as an anthropomorphic conflict, in which each object (or person) has "its force," "its dynamism," then how is it possible to accept that one of the two can win without exerting on the other a force of greater magnitude? The huge resistance observed in these cases, even among authors of textbooks,[11] is significant. In terms of common reasoning, such resistance is a major phenomenon.

A revision of the usual teaching objectives is called for; until now, Newton's "third law" has been simply ignored, and any amount of heresy on this subject has been tacitly tolerated. Our analysis recommends specific educational approaches, starting with a schematisation procedure that avoids the usual ambiguity. It allows teachers to analyse what physical examples can be proposed to pupils: situations involving acceleration are difficult, but preferable to those involving equilibrium, which are often uninteresting. Indeed, they do not teach anything of use, as they cannot clear up the ambiguities we have pointed out (box 6, column c). Also, contrary to popular belief, situations of interaction at a distance are, in a way, easier to deal with than situations of contact, because the absence of a point of contact obliges one to take a stand and really decide on which object a force is supposed to be exerted.

[11] See Menigaux (1986) and Viennot (1979a, b, 1982a, 1989a).

Box 5

Separating objects to clarify analysis – See Viennot (1989a)

Proposal

"Fragment" diagrams – i.e., depict objects as separate even if, in fact, they are in contact with one another.

Aim

To put an end to traditional diagrams that "encourage" common forms of reasoning, by not showing on which object each force is acting.

Examples

A driver whose car has broken down is pushing it.

Common diagram and comments | **Proposed diagram** ("fragmented" diagram)

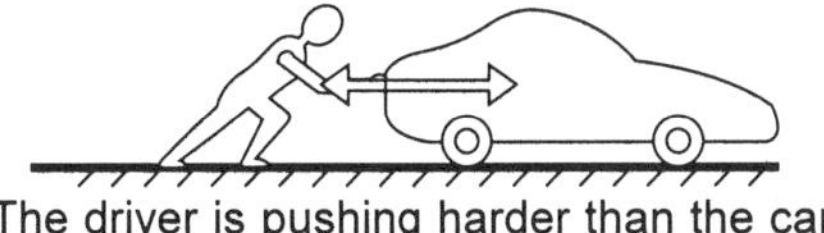

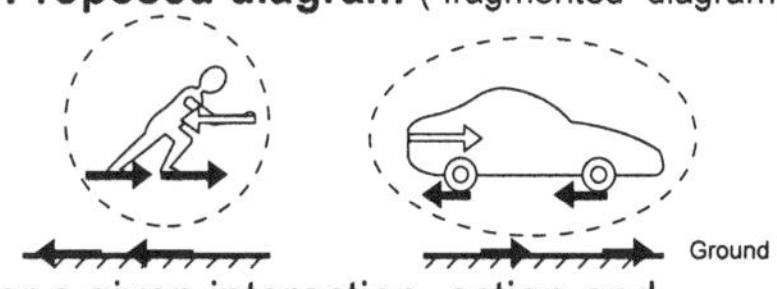

"The driver is pushing harder than the car is resisting: action is larger than reaction."

For a given interaction, action and reaction always have the same intensity.

A wrestler defeats his opponent

Common diagram and comments | **Proposed diagram** ("fragmented" diagram)

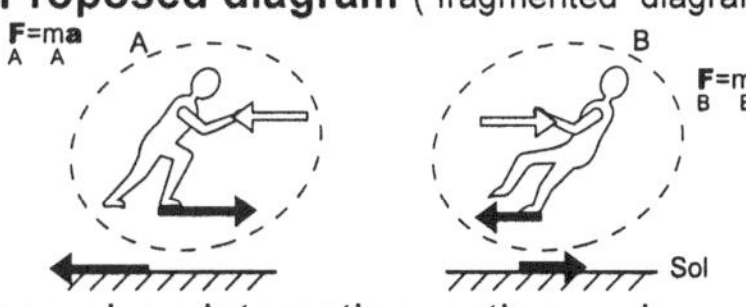

"A is pushing harder than B, action is larger than reaction."

For a given interaction, action and reaction always have the same magnitude.

Important details

- List the objects that must be separated and those for which we want to add up the forces acting.
- List the interactions, giving each one a colour.
- If we are concerned only with the overalll displacement of the objects, it is not necessary to worry about the exact point of application of each force on a given object (see chapter 10, box 1).
- We can also neglect certain interactions whilst working on others – deal with only one direction at a time (e.g., horizontal interactions).
- Two arrows of the same colour must
  a) point in opposite directions
  b) be the same length
  c) be in different "bubbles"
- A "bubble" contains one arrow for each interaction considered. For objects at the end of the chain of analysis (here, for example, the ground), which are incompletely analysed, the sum of all the forces acting cannot be established.

Box 6

Clarifying the analysis of an object and its support

It is often thought that an object "exerts its weight on its support". This is only true at equilibrium (and even then, the wording is questionable: see below).

Example: type a. diagrams are often used to depict an oscillating mass (attached to a spring) accelerating upwards.

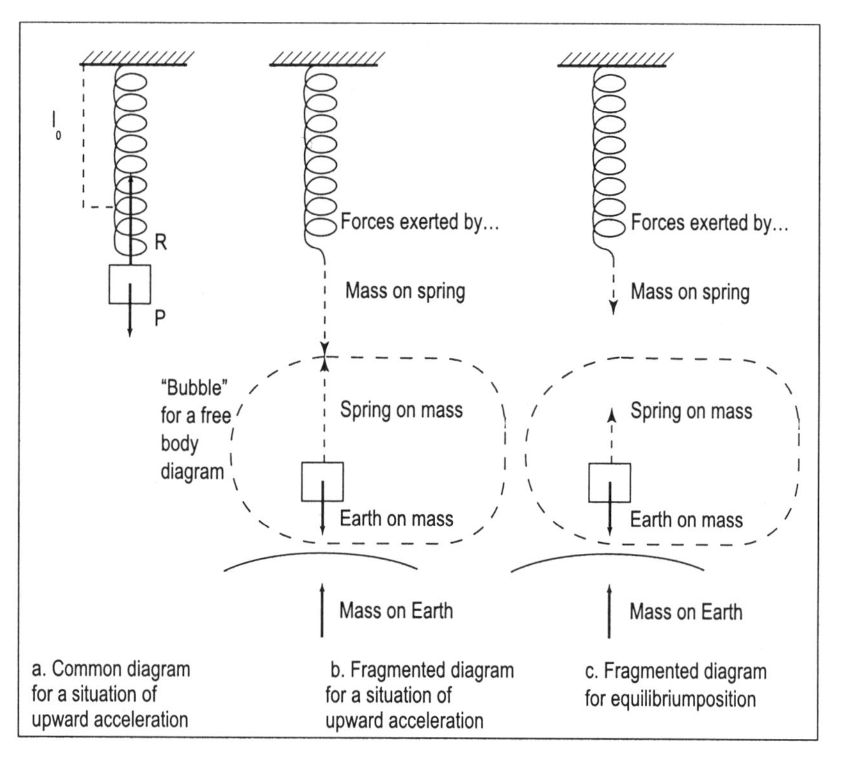

a. Common diagram for a situation of upward acceleration

b. Fragmented diagram for a situation of upward acceleration

c. Fragmented diagram for equilibriumposition

Diagram a., though not really wrong, is compatible with the reasoning below:

Weight of M = force of M = force of (exerted by) M on the spring = action of M

then

Weight of M < Force of the spring on M

therefore

Action < Reaction

which leads to the conclusion frequently found in students' comments:

The law of action and reaction no longer applies.

The "fragmented diagrams", b. and c., show that

- the mass does not exert its weight on the spring. Moreover, the mass-spring interaction is zero when, during oscillation, the length of the spring is the same as when it has no load.
- at equilibrium, the mass-spring interaction is of the same value as weight is, but it is not weight.
- the law of reciprocal actions applies in all these situations.

Box 7
Clarifying the analysis of centrifugal force – See Viennot (1989 a)

Teachers often say that centrifugal force is a "fictitious force", meaning that it only arises when the problem is considered in the frame of reference of the moving object.

- But we can "feel" it, for example...

when we spin a stone horizontally at the end of a string or when we are pushed against the car door as the car turns

- Correct analysis in a Galilean frame of reference
  Only horizontal forces and motions are analysed

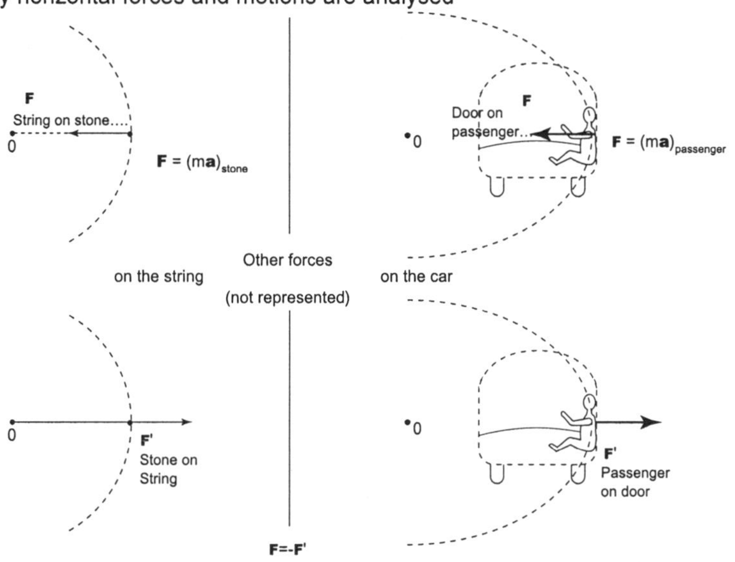

View from above for horizontal motion   View from behind in a horizontal turn

A non-fictitious centrifugal force is exerted on whatever makes the object turn.

- Common types of diagrams for these situations:

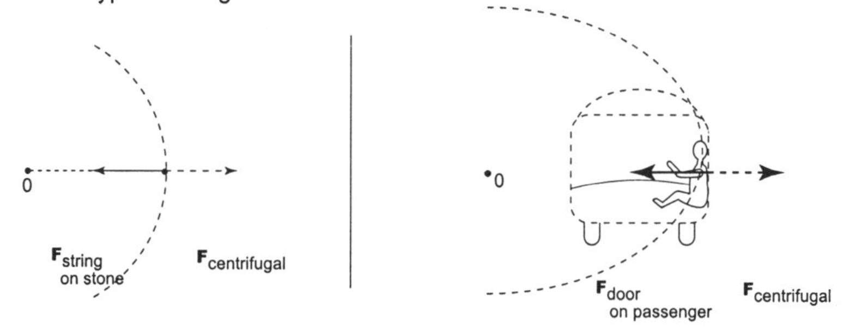

- They contribute to the vagueness about the use of the term centrifugal force; moreover, the usual wording is ambiguous: "centrifugal force **of** the stone", "...**of** the passenger."
- They cannot be interpreted by changing the frame of reference, since the trajectory that is represented is not reduced to a single point.

The schematisation procedure that we suggest can be applied to situations that are both common and the source of frequent error (appendix 3). It makes it clear that if a nail can be driven into a board, or if a sled makes tracks in the snow, it is not because "action is larger than reaction". In the analysis of friction (box 5), the use of fragmented diagrams seems essential to counter the idea that friction always opposes motion.[12] This technique is also helpful to explain that Archimedes' upthrust has a reciprocal effect: a downward force is exerted on the fluid by the immersed body. And it can also make it clear that an object does not exert "its weight on the support" (box 6), except in situations of equilibrium: and even then, the magnitudes are equal, but the interactions are not identical. As for circular motion, the procedure suggested here clarifies matters (box 7): a force can be centrifugal without being "fictitious." Indeed, the force exerted by an object on what constrains it to a circular trajectory is directed radially towards the outside of the circle, even in a frame of reference that is as Galilean as can be.

Such ideas were rarely voiced before the didactic studies cited here were conducted.

5. THE STAKES IN TEACHING DYNAMICS

This chapter shows that natural reasoning is in fact coherent. Faced with moving objects and signals, when motion seems inexplicable, natural reasoning uses its most readily available resource – the idea of an object storing within it a previous cause of movement, in the form of a vague dynamic notion: "the force of…."

It is quite a job to "disentangle" the causal content which is natural in spontaneous analysis into various identified concepts – force, velocity, linear momentum, or energy.

In particular, in correct physics, the association of force and velocity has got to be avoided; the first quantity must be associated with the rate of change of the second. Simplified diagrams such as those that appear in box 1 help to study situations in which motions differ but the acceleration is the same.

Granted, this concern is not really new, but the coherence of natural thought must now be taken into account: the aspects of reasoning that emerge form a whole. The attribution of a force to the object and temporal gaps between quantities are two of many elements of reasoning that demand more general consideration, or we run the risk of being ineffective.

[12] See chapter 3, box 3; also Caldas and Saltiel (1995).

We should be especially cautious when using the ubiquitous word "of". And we should stop considering that the location in time of the quantities that we deal with is obvious, under the pretext that, in fact, the different terms of a relationship are all measured at the same time. This simultaneity is a difficulty. It must be made explicit in class. Once again, using diagrams like those suggested above (box 5) is extremely helpful. Then the two Newtonian laws discussed here can appear as "separate" from one another. There should no longer be any confusion between an evaluation of the balance of forces on an object and the analysis of the reciprocal actions of two objects.

On all these points, the test questions proposed in this chapter can serve as aids in pedagogical activities involving the recognition of difficulties, debates with the pupils, and new theoretical framing (Saltiel and Viennot, 1983; Viennot, 1985, Maurines and Saltiel, 1988b).

But, at the same time, we must do more than lift the law of reciprocal actions out of the rut of anthropomorphic conflict, or instil in students a more acute sense of the passage of time.

The series of propositions contained here should help to establish scientific thought as a viable alternative to natural ways of thinking, one that offers greater coherence and reliability.

Obviously, this raises questions regarding educational objectives and the extent to which they can be attained. Some documents presented in appendix 4 show that such concerns have been given much thought in France since the early nineties. The choices made in 1993 concerning grades 9 and 11 (science section)[13] are in line with the conclusions presented in this chapter. They consist in presenting Newton's laws in summarised form, but uncompromisingly. The laws are taught in summarised form, because advanced kinematics, which demands complex analytical expression, is avoided. At grade 11, for instance, it is enough to define zero acceleration (and, conversely, non-zero acceleration) and to associate it with a resultant force equal to zero (and, conversely, not equal to zero). This is done uncompromisingly, because difficulties are pointed out and dealt with at once.

Again, at grade 9, frictional propulsion is introduced – a provocative expression, which calls for a precise analysis of the pair of forces involved in the interaction.

Increasing our efforts towards establishing the coherence particular to physics and showing that its logic surpasses that of natural reasoning may well entail less emphasis on the manipulation of formulae. In this context, "qualitative" is clearly not synonymous with "sloppy."

[13]Troisième and Première scientifique: respectively, the fourth and sixth year of secondary education in France.

APPENDIX 1

A CAUSE SITUATED IN THE PAST: PROPULSION BY A SPRING

For more detail, see Viennot, 1979.

When an object persists in rising even though gravity should in all logic pull it downwards, common reasoning attributes this to a cause: a dynamic (and undifferentiated) "capital" stored within the object.

Taking the case of a mass placed on a spring, then pushed downwards and released, we can ask, how far would one need to push the mass down for it to rise into the air when the spring is released?

One condition often suggested by the students (box 8) is that "The force exerted on the mass by the spring at maximum compression must be greater than the weight of the mass." Box 8 gives the balance of forces at the critical stages in the situation, and the solution to the problem. The condition proposed by the students is inevitably realised as soon as the mass reaches a point lower than its starting point, and is not, therefore, sufficient to ensure that the mass will rise off the spring: the mass may oscillate around its equilibrium position without rising. Moreover, at the moment the mass rises, the force, which, according to the students, should be greater than the weight, is in fact zero: the spring is free, neither stretched, nor compressed (its own mass is not sufficient to stretch it considerably due to inertia). The upward movement of the mass cannot be explained simply in terms of a balance of forces. Once again, the only force acting on the projectile is weight, hopelessly directed downwards. This is outrageous: a cause must be found. And so it is looked for in the past; a balance is established between two forces that exist at different moments in time (poor Newton!); and, often, the satisfying conclusion is that "the law of action and reaction no longer applies."

Box 8
Understanding propulsion by a spring (Viennot 1979)

Outline of the question

A mass is set on a spring. It is not attached to it. How far down would one need to push the mass for it to rise into the air when the spring is released?
(spring constant: k)

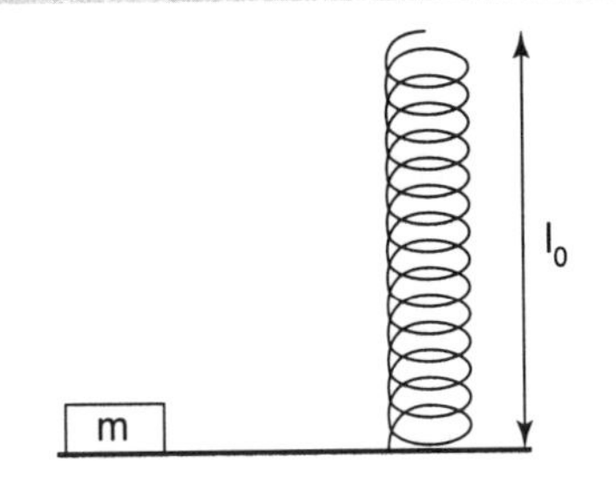

Outline of the correct solution

Let $E_p(z)$ be the potential energy of the Mass-Spring-Earth system. If $E_p(z=\perp_0)=0$, one gets the following expressions and curve:

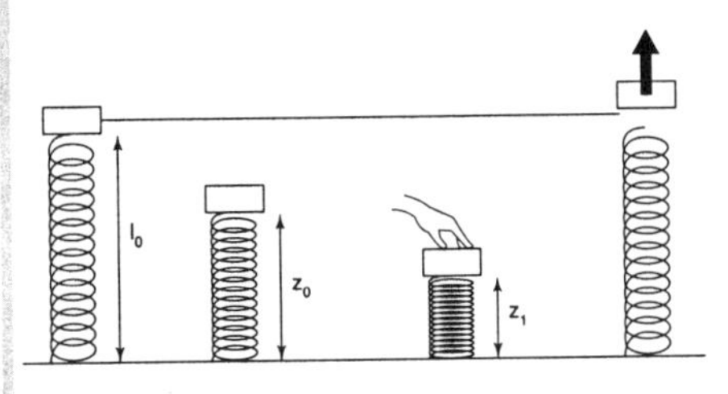

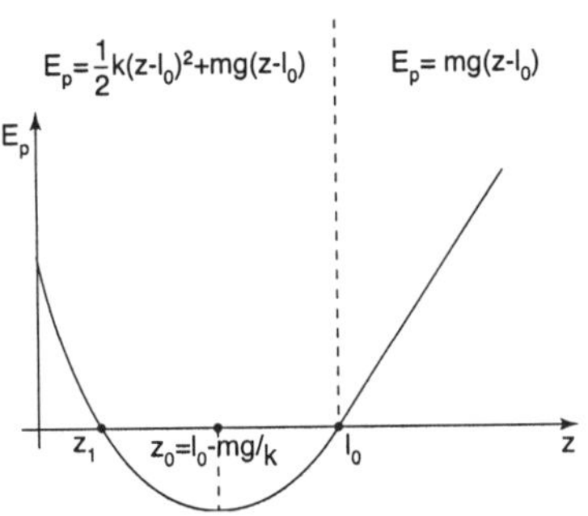

Force diagrams (fragmented [a])

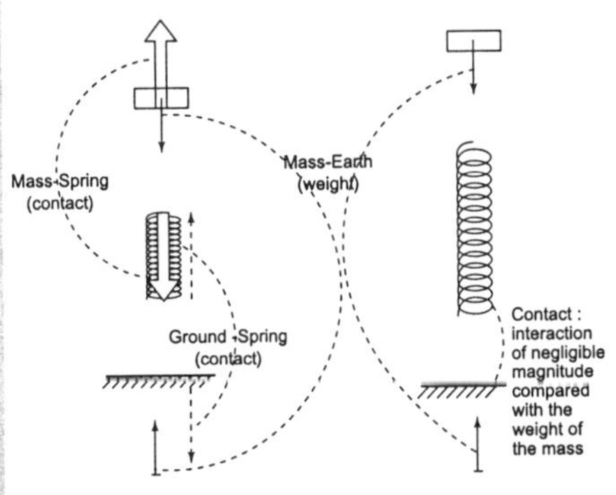

Conditions for the mass to rise

For the mass to rise off the spring, it must reach an altitude of at least l_0.

The mass must therefore push the spring down to an altitude of at least z_1, where the potential energy E_p has the same value as at l_0:

$$E_p(z_1)=0$$

The condition for the mass to rise is therefore:

$$\frac{1}{2}k(z_1-l_0)^2+mg(z_0-l_0)=0$$

Common answers

- Most are structured as follows:

Force or energy of the spring > Force or energy of the mass

- They deny the law of reciprocal actions: "For the mass to take off, the law of action and reaction must cease to be true."

a. The points of application of the forces are not specified. Only the **object** that is subjected to the force matters here.

APPENDIX 2

WHEN THE PAST LEAVES ITS MARK

In certain cases, the dynamic "capital" of a moving object is not understood as dissipating immediately, as soon as friction disappears, as the following questionnaire shows:

You are in a plane flying horizontally. You put an ice-cube from your whisky on a perfectly smooth (frictionless) horizontal table. At time t_0 you release this ice-cube without adding linear momentum.

Will it move in any direction in either of the following cases? If yes, which one? Explain.

The plane is...

1 ...making a turn at constant speed

YES NO

2 ...beginning to turn at time t_0, at constant speed

YES NO

In each case, draw the trajectory of the ice cube

a) on the table, in the plane

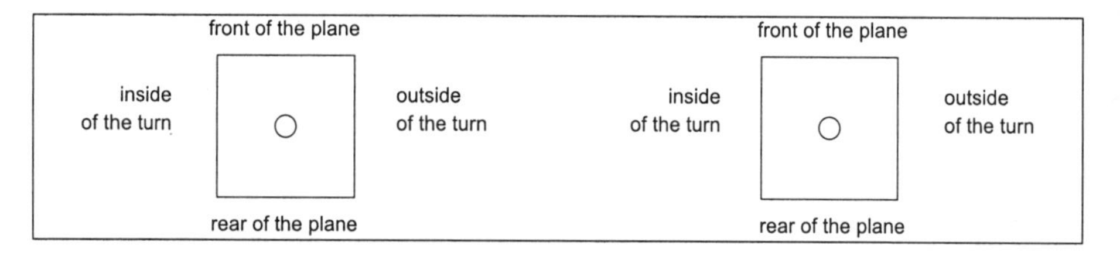

b) in the sky, imagining that the plane and the table are invisible, and only the ice cube is visible.

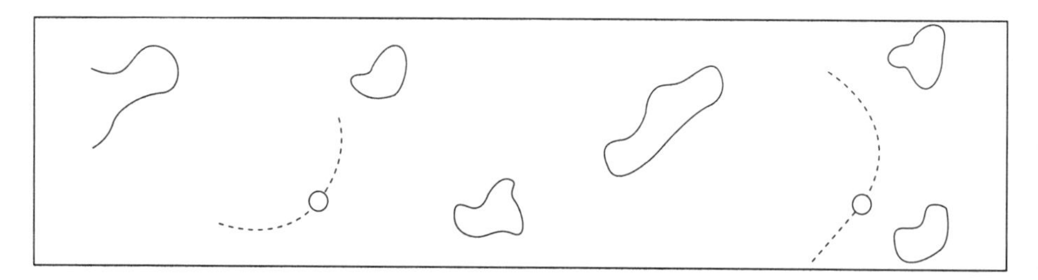

NB:

The text is not realistic: a plane making a turn tilts. But this does not explain the characteristics found in students' answers to the question. The author wanted to distinguish as clearly as possible between two frames of reference: that of the plane and that of "the sky". Some may prefer a situation in which a limousine makes a turn on a (practically) horizontal surface. However, the types of answers obtained for this questionnaire cannot be explained by the lack of realism of the situation proposed.

Only a starting direction is asked for, not a complete trajectory.

OUTLINE OF CORRECT ANSWERS

Any kinematic data before time t_0 is not pertinent. Only the velocity of the ice cube, the forces acting on it, and the subsequent motion of the plane matter here. These elements are the same for questions 1 and 2. The answers to both questions are therefore identical.

The motion of the ice cube in the (Galilean) reference frame of the sky obeys the following relationship:

F exerted on the ice cube = m**a** of the ice cube

The resulting force **F** exerted on the ice cube is zero everywhere, and, therefore, so is the acceleration of the ice cube. The latter continues in its course, tangential to the initial trajectory, at the same velocity. The plane does not. The diagrams below correspond to three different moments t_0, t_1, t_2, and show that the ice cube moves relative to the table.

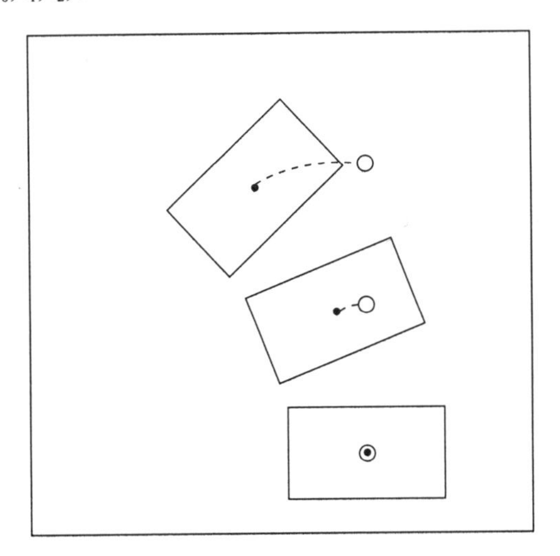

The initial directions of the motion of the ice cube can therefore be represented as follows:

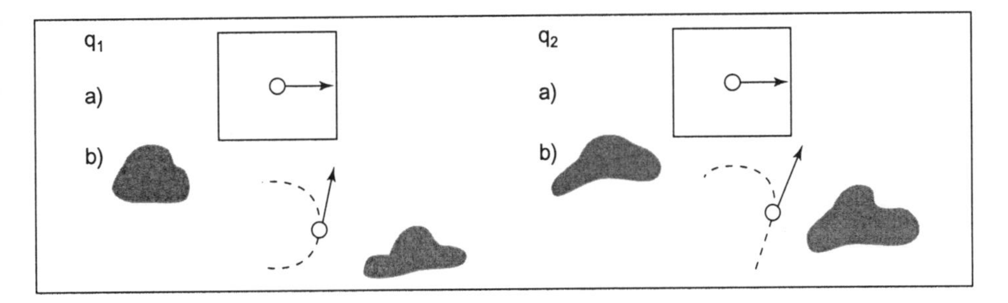

FREQUENT ANSWERS

The answers are very mixed. But when the ideas that informed the design of the question are considered, the picture becomes clearer.

The two things to be assessed were:

- whether the students realise that the answer in no way depends on the past history of the motion, but only on the velocity of the ice cube at the moment it is released, and on what happens after that;
- whether, for each question, the students propose different trajectories according to the frame of reference.

The student who gave the answer below has formed an opinion on each point. According to him,

- the past history of the motion does count;
- the directions of the motions are transposed from one frame of reference to the next: the ice cube starts "radially" (question 1) or else "tangentially" or "forwards" (question 2), in both frames.

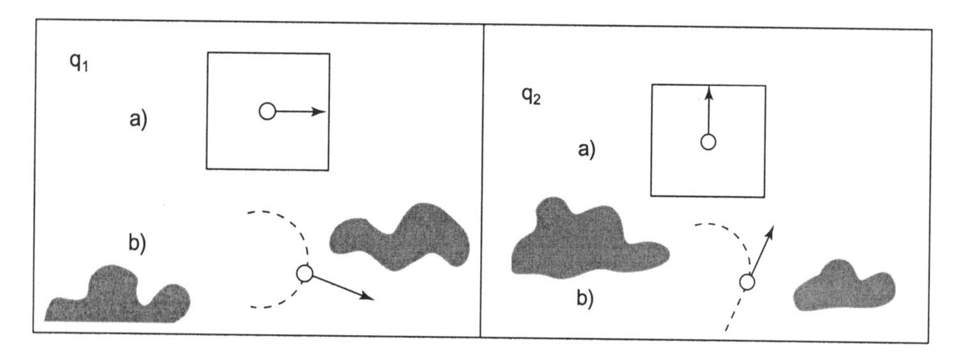

These examples illustrate the same aspects of answers, but with more complex trajectories.

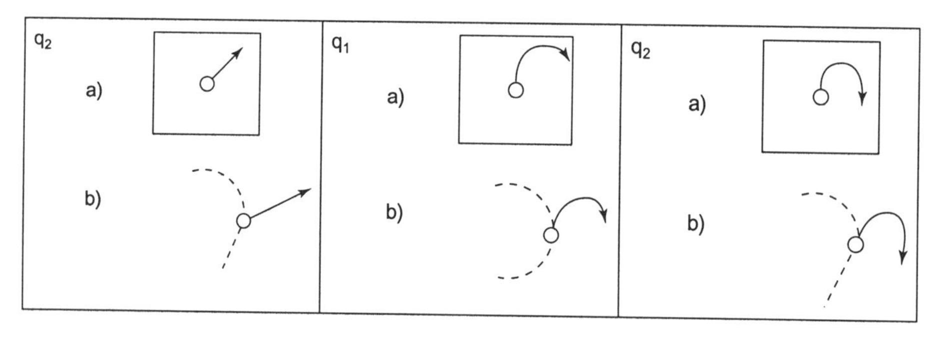

These aspects account for a large proportion of answers, as the following table shows:

	Different answers for Q 1 and Q 2	Answer for Q 1 or Q 2 transposed from one frame of reference to another.	Correct answers for Q1 and Q2
First university year (N=79)	34%	87%	4%
Mathématiques Supérieures [a]	29%	68%	15%
Mathématiques Spéciales [a]	b	80%	10%

a. In France, Mathématiques Supérieures and Mathématiques Spéciales are respectively the first and second years of preparation for the competitive examinations for admission to institutions (*grandes écoles*) specialising in science.
b. No data available

The answers combine two previously identified elements of common thought:

- The tendency to "rigidify" trajectories and directions results in their being understood as independent of a frame of reference.

- Although a moving object has no recollection of the past beyond its present velocity, it is seen as keeping a more "informed" trace of it stored inside itself: a "supply of cause," so to speak. Here, this trace is the type of motion just depicted (rectilinear or circular).

APPENDIX 3

ANALYSING INTERACTIONS: TWO SITUATIONS

Viennot, 1989

A nail is hammered in

Common diagram and comments

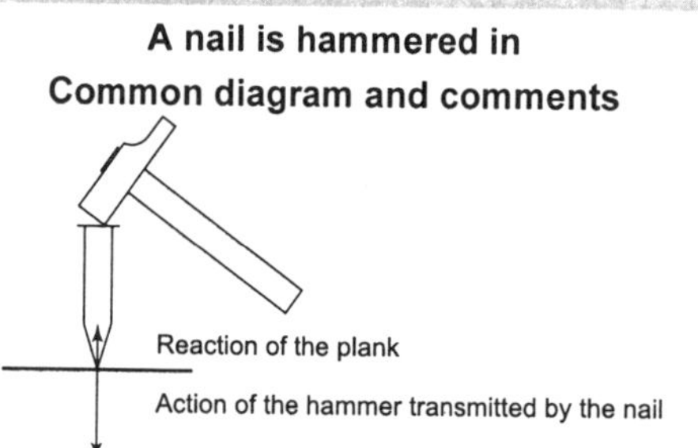

Action of the nail > reaction of the plank

Archimedes' push: does it work downwards?

Question:

Does the marker on the scale move if the ball is removed from the water?

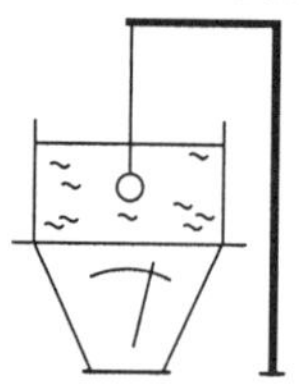

Common answers:

"The marker does not move"

"Water exerts its weight on its support"

Archimedes' buoyant force is considered as having no reciprocal force.

Outlines of correct answers

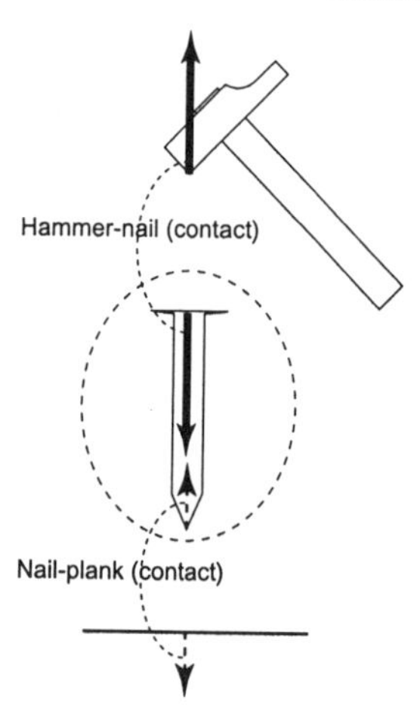

The balance relating to the nail when it accelerates shows that the hammer-nail interaction is of greater value than the nail-plank interaction.
The law of reciprocal actions still applies.

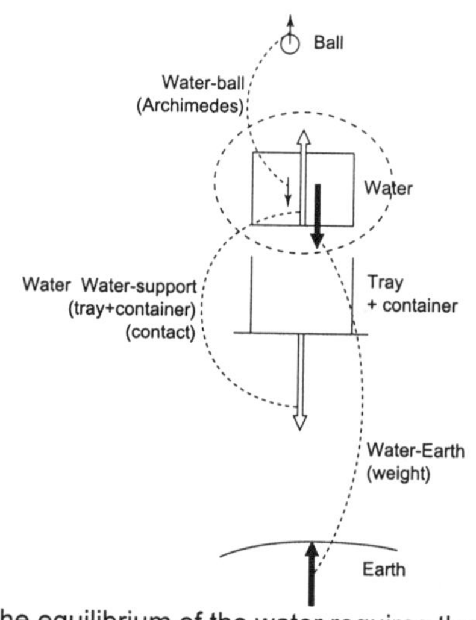

The equilibrium of the water requires that the action of the water on the support be of greater value than that of its weight. Therefore, the marker is at a higher setting than when the ball is removed.

Interactions concerning the nail, negligible compared to those mentioned previously:

- Nail-earth (gravitation)
- Air-nail (Archimedes' push from the air)

Interactions that are not considered:

- Other interactions concerning the hammer and the ground

Interactions concerning water, negligible compared to those mentioned previously:

- Unit (water-container) – Air (Archimedes' push from the air)

Interactions that are not considered:

- Other interactions concerning the ball, the support, the scale, the ground, the earth

For each object, no particular attention is given to the exact point of application of the forces because only the motion of the centre of inertia is analysed (see also chapter 10, box 1).

APPENDIX 4

EXCERPTS FROM OFFICIAL INSTRUCTIONS ON THE CURRICULUM FOR GRADES 9[14] AND 11 (SCIENCE SECTION)[15]

Aspects of these texts that particularly correspond to the contents of the preceding chapter have been italicised by the author.

Excerpts from the official comments on the grade 9 curriculum
(Bulletin Officiel, 1993)

"Propulsion and means of transport"

1 Propulsion and braking

1.1 This curriculum should not be confused in any way with the previous curriculum for grade 10,[16] which posed great difficulties as regards linear momentum and its conservation.

A highly phenomenological approach to real situations (...) is encouraged. This approach leads to *a first representation on a modest scale; vectors need not be calculated: only diagrams are used, which take into account the direction of forces and their comparative magnitudes (greater than, smaller than, equal).*

The same situations can be used to define the characteristics of a given motion and to identify the various forces involved. In this way, the effects of a force may be identified.

1.2 As has been pointed out earlier, the aim is to start from simple or slightly more complex situations in order to draw up a preliminary inventory of forces, in the following manner:

Action of object... on object....

Or: Object... acts upon object....

Or: Object... pulls (pushes) object....

In the course of this preliminary inventory, emphasis is placed on the fact that, even at a distance, certain objects interact (through magnetic, electric, or gravitational interactions).

Modelling the action is the next step. The previous examples are used to associate each action with a force whose direction and magnitude (in newtons) are specified by means of a vector.

The example of free fall illustrates *the effect of a force on the motion of an object: the value of the velocity increases or decreases, the trajectory curves in the direction of the force...* In particular, it illustrates the fact that *the direction of the force is not necessarily the direction of the motion.*

This is designed to train pupils to be rigorous in their analyses. Analysis is made as concrete as possible through the use of diagrams that methodically represent the objects under study and the forces acting upon each one. Particular attention is given to contact forces, whose representation is difficult. Even if it is not stated among the objectives, it is necessary

[14] Troisième, i.e., the fourth year of secondary school in France.

[15] Première scientifique, i.e, the sixth year of secondary school in France; science section.

[16] Seconde, i.e., the fifth year of secondary school in France.

to distinguish between local and global actions. *As long as it is the motion of the whole that is being considered (in fact, the motion of its centre of mass), it is not necessary to discuss the point of application.* If, on the other hand, the rotation or deformation of the object are considered, then the problem cannot be avoided.

(...)

1.3 The key idea in this paragraph is that *the friction force between the ground and the moving object is necessary to propulsion.* This idea must be introduced through experimentation.

(...)

Excerpts from official comments on the grade 11 curriculum
(Bulletin Officiel, 1992b)

1.3 Interactions between objects

Various simple situations are described in terms of interactions, stress being laid on the importance of contact actions and friction phenomena: contact between a tyre and the road, the action of the wind, upthrust, etc. At this point it can be explained, for example, that *it is really friction that makes motion on the ground possible.* It must be pointed out that contact actions are generally repulsive, but one might bring up the fact that contact can be violently attractive, as it is in the case of two perfectly smooth surfaces in contact with one another, and adhesion techniques can be mentioned. (...)

The *principle of "reciprocal interactions"* or the principle of action and reaction" is in fact Newton's third law. *It constitutes a major conceptual difficulty. It is, for example, difficult to admit that even if a man is not touching the ground, he is exerting upon the Earth a force of the same magnitude as his weight; or that a body immersed in a liquid exerts upon it a force that is the opposite to the upthrust. Nor is it easy to accept that when one pushes down on an object, the force it exerts in return is the opposite of the force exerted upon it. Rigorous schematisation aids in establishing a distinction between forces: the forces exerted on a given object, which are considered when adding together the forces acting on that object (with a view to applying the principle of inertia and the fundamental relationship of dynamics at a later time), and the two forces involved in the interaction, which are exerted on two different objects and are always opposite forces. (...)*

Even though formalism is not introduced until grade 11, it is important to illustrate with concrete examples or experiments the consequences of $\Sigma\mathbf{F}=0$ on the motion of the centre of mass. *In particular one needs to start eradicating the fallacious but widespread notion that* $v=0$ *at time t implies* $\Sigma\mathbf{F}=0$, *and that* $\mathbf{v}\neq 0$ *implies* $\Sigma\mathbf{F}\neq 0$.

Moreover, the "Accompanying Documents" (1992, 1993) established by the Technical Group (GTD) for Physics, that formulated the curriculum proposals and comments, propose two versions of the schematisation procedures described above (fragmented diagrams) for grades 9 and 11.

REFERENCES

Andersson, B. 1986. The experiential Gestalt of Causation: a common core to pupils' preconceptions in science. *European Journal of Science Education*, 8 (2), pp 155-171.

Bulletin Officiel du Ministère de l'Education Nationale, 1993, n°93, *Nouveaux programmes de physique et chimie pour la classe de Troisième des collèges*, pp 3721-3737.

Bulletin Officiel du Ministère de l'Education Nationale, 1992b. *Nouveaux programmes de physique et chimie pour les classes de Seconde, Première, et Terminale des lycées, Numéro hors série du 24-9-1992*, Vol II, p 38.

Caldas, E. 1994. *Le frottement solide sec: le frottement de glissement et de non glissement. Etude des difficultés des étudiants et analyse de manuels.*. Thesis. Université Paris 7.

Caldas, E.and Saltiel, E. 1995. Le frottement cinétique: analyse des raisonnements des étudiants. *Didaskalia*, 6, pp 55-71.

Driver, R., Guesne, E. and Tiberghien, A. 1985. Some features of Children's Ideas and their Implications for Teaching, in Driver, R., Guesne, E. et Tiberghien, A. (eds): *Children's Ideas in Science*. Open University Press, Milton Keynes, pp 193-201.

Groupe Technique Disciplinaire de Physique 1993. *Document d'accompagnement du programme de Troisième*. Ministère de L'Education Nationale, Paris.

Groupe Technique Disciplinaire de Physique 1992. *Document d'accompagnement du programme de Première*. Ministère de L'Education Nationale, Paris.

Gutierrez, R. and Ogborn, J. 1992. A causal framework for analysing alternative conceptions, *International Journal of Science Education.* 14 (2), pp 201-220.

McDermott, L.C. 1984. Revue critique de la recherche dans le domaine de la mécanique. *Recherche en Didactique: les actes du premier atelier international, La Londe les Maures, 1993.* CNRS, Paris, pp.137-182.

Maurines, L. 1986. *Premières notions sur la propagation des signaux mécaniques: étude des difficultés des étudiants*. Thesis. Université Paris 7.

Maurines, L. 1993. Mécanique spontanée du son. *Trema*. IUFM de Montpellier, pp 77-91.

Maurines, L. and Saltiel, E. 1988a. Mécanique spontanée du signal. *Bulletin de l'Union des Physiciens*, 707, pp 1023-1041.

Maurines, L. and Saltiel, E. 1988b. *Questionnaires de travail sur la propagation d'un signal*, Université Paris 7 (diffusion LDPES)

Menigaux, J. 1986. Analyse des interactions en classe de troisième. .*Bulletin de l'Union des Physiciens*, 683, pp 761-778.

Saltiel, E. and Viennot, L. 1983. *Questionnaires pour comprendre*, Université Paris 7 (diffusion LDPES).

Saltiel, E. and Viennot, L. 1985. What do we learn from similarities between historical ideas and the spontaneous reasoning of students? *The many faces of teaching and learning mechanics*. In Lijnse, P. ed.. GIREP/SVO/UNESCO, pp 199-214.

Séré, M.G. 1982. *A propos de quelques expériences sur les gaz: étude des schèmes mécaniques mis en oeuvre par les enfants de 11 à 13 ans,* Revue Française de Pédagogie, 60, pp 43-49.

Séré, M.G. 1985. *Analyse des conceptions de l'état gazeux qu'ont les enfants de 11 à 13 ans, en liaison avec la notion de pression, et propositions de stratégies pédagogiques pour en faciliter l'évolution.* Thesis (Doctorat d'état). Université Paris 6.

Viennot, L. 1979. *Le raisonnement spontané en dynamique élémentaire*, Hermann, Paris.
Viennot, L. 1979 Spontaneous Reasoning in Elementary Dynamics, *European Journal of Science Education*, 2, pp 206-221.
Viennot, L. 1982a. L'action est-elle bien égale (et opposée) à la réaction?, *Bulletin de l'Union des Physiciens*. n° 640, pp 479-488.
Viennot, L. 1985. *Mécanique et énergie pour débutants*, Université Paris 7 (LDPES).
Viennot, L. 1989a. Bilans de forces et lois des actions réciproques. *Bulletin de l'Union des Physiciens*. 716, pp 951-970.

Chapter 5

Quasistatic or causal changes in systems

In association with Jean-Louis Closset and Sylvie Rozier

To draw out and highlight the basic elements of physics and "natural" reasoning, our analysis has, up to now, centred on changes in relatively simple objects. But the complexities of nature and the possibilities of physics are greater than those we have described, even if we remain within the realm of the "elementary".

1. THE ESSENTIAL: SYSTEMS THAT OBEY SIMPLE LAWS

Complex groups of elements in mutual interaction – "systems" – can often be described relatively simply by means of a few laws.

What simplifies description is the hypothesis that these elements mutually inform one another very quickly, relative to the time it typically takes for the whole group to change. It is often said, for brevity's sake, that with these systems one can "neglect internal propagation," which means, more precisely, that one can neglect its duration. But this is an approximation – one that allows certain laws to "hold" despite, and during, a change in the system. One might call them "quasilegal" evolutions of systems.

Textbooks use the adjectives "quasistationary" (in connection with electricity, for example) or "quasistatic" (in connection with thermodynamics). But there is one field in which, most of the time, no one remembers to mention these types of change: elementary mechanics. The particles of a solid mutually inform one another very quickly when anything occurs (if the reader will forgive this anthropomorphic image). It therefore seems unnecessary to specify that "one can neglect internal propagation of effects". Nevertheless, the transfer of information sometimes takes an amount of time that is not negligible relative to the time scale of characteristic events: there can be waves in solids, too.

Box 1 illustrates some simple examples of quasistatic analyses taken from the fields of mechanics and electricity.

The notes accompanying each example show that many quantities are involved: situations of this sort are generally called multi-variable, or multifunctional, problems. Here, complexity is reduced, but not eliminated.

Then there are the laws. Some describe, in a phenomenological fashion, each of the parts (the "subsystems") involved. Others express relationships between subsystems – e.g., those expressing conservation, or a fundamental law, such as the law of reciprocal actions.

Box 1

Quantities and relationships in a quasistatic analysis of systems (examples)

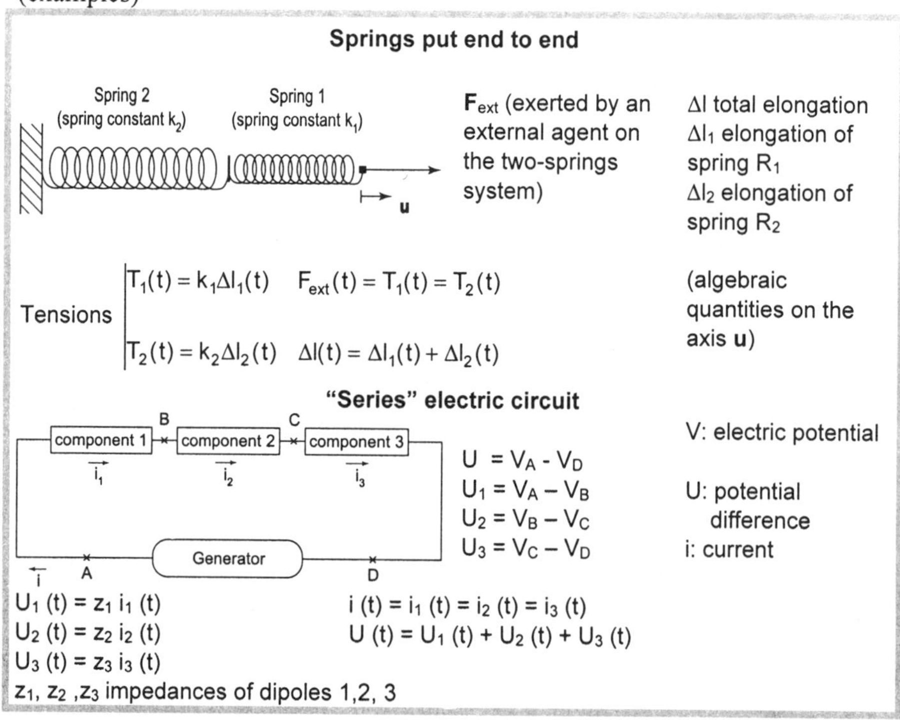

The general practice is to make no mention of time. But the remarks in chapter 4 apply even more strongly here. Many of the quantities involved are modified in the course of the transformation; everything can "change" at the same time, since everything is "informed" at the same time. Here, again, events are simultaneous.

In this type of "quasilegal" description, the laws remain unchanged throughout the process of "multiple-transformations." They can therefore be written in more explicit form, with a reference to time alongside each physical quantity (see box 1).

2. NATURAL REASONING: MORE STORIES

Thinking in terms of simultaneity does not come naturally to us, as has been shown. Moreover, the idea of an object is ever-present in natural thought. It is logical that these elements of common reasoning should crop up again here. And they do: an analysis of the evolution of systems is very often structured like a story, in the sense that there is a temporal link between events. Although in a quasistatic analysis one should think in terms of "at the same time," common reasoning prefers the storyteller's "later on," and, sometimes, "further on."

The idea of something happening "further on" presupposes that there is a clear spatial structure in the system, which is not always the case. In the first examples discussed below, the situations are ones where there is a "natural path" in space to guide reasoning; the subsequent examples show that the story can also take place "on the spot."

3. SYSTEMS WITH A CLEAR SPATIAL STRUCTURE

3.1 Springs connected end to end

Let us imagine a "system" made up of two springs, R_1 and R_2, with length l_{01} and l_{02}, and spring constants k_1 and k_2. These springs are connected end to end, as shown in box 2. The variables characterising each spring are given there, as well as their interactions, both with each other and with the environment. Some describe the state of each spring (extensions Δl_1 and Δl_2, the sum of which gives the total elongation Δl) and others the forces involved ($\mathbf{T}_1$: magnitude of the force acting on the lower end of spring R_1; $\mathbf{T}_2$: the same for R_2; $\mathbf{F}_{ext}$: magnitude of the force exerted by the experimenter on the lower end of the set of two springs). These quantities change simultaneously during a transformation of the system, if it is stretched, for example. But the relationships given in box 1 remain valid at every moment of the transformation. This implies something that is not always obvious at first: the transformation is quasistatic; all of the elements of the system are informed about what is taking place at the same time. In short, "one neglects internal propagation".

Each state of the system can be described as if it were fixed, without consideration of what brought it to that point or of what will follow: in that sense time does not enter into this problem.

This situation was presented in a survey questionnaire (Fauconnet, 1981) to evaluate the impact of the temporal aspect of events on the reasoning of students. One version (box 2) proposes a comparison between two systems, both made up of the same elements – two springs that are identical to R_1 and R_2 respectively but which are in different **states** of elongation: the first system is free, the second is stretched. The second version (box 3) proposes another system made up of elements that are identical to R_1 and R_2, but this time they have undergone a **transformation**: the system is being pulled on. In both versions, the same data is given, i.e., the total elongation in the "stretched" state, and the student is asked to find the displacement of the junction point and the external force exerted in this state.

Box 2

Two analyses of a two-spring system (from Fauconnet, 1981)

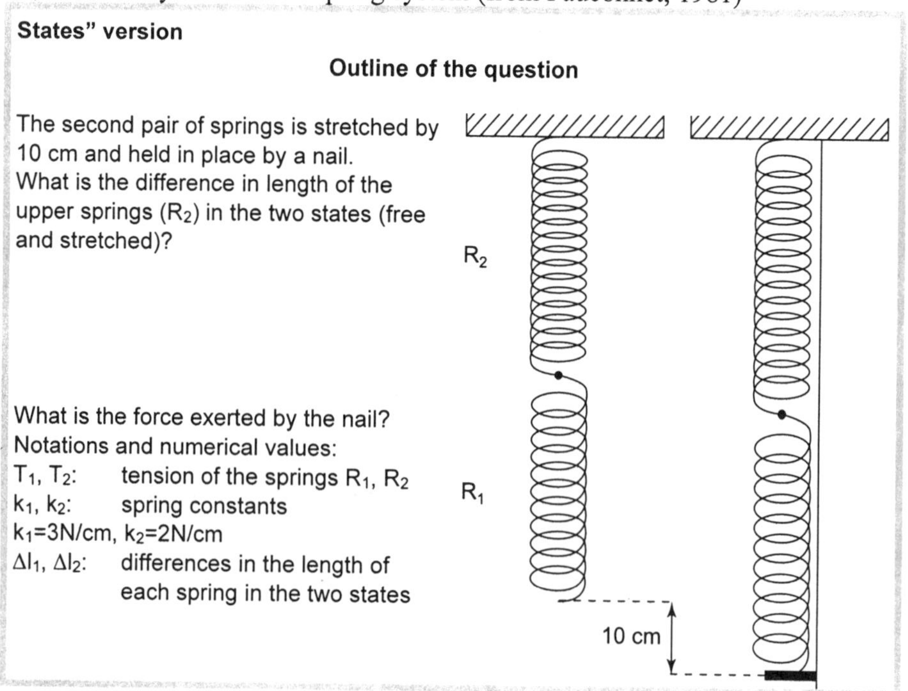

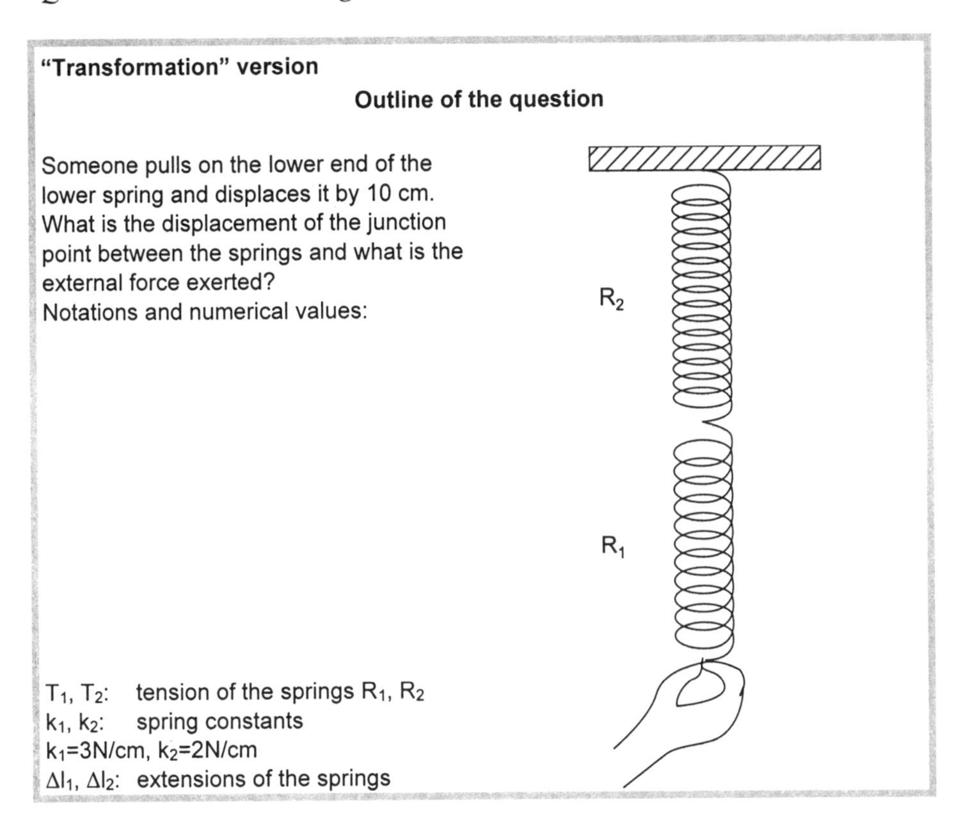

The quantities involved and the relationships between them are exactly the same in both versions of the question ("states" or "transformation"). Do the students' responses clearly reflect this similarity?

Each participant was only given one version of the question. So only the frequencies of characteristic responses in groups that dealt with one of the two proposed situations can be compared.

The students' diagrams (box3) show that the force F_{ext} exerted by the external agent is "opposed," in a sort of "balance," by the two other forces, each acting upon one of the springs. Newtonian physics tells us that such a balance cannot provide information about the acceleration of a given body ("a system"), since the forces acting are not all external to the system being considered. But the diagrams indicate that students envisage this situation (the author speaks of how they "read" the situation) in the by-now familiar terms of an anthropomorphic conflict: each element of the two-spring system resists in its own way the aggressive action of the external agent – using "its own force" (even though this means ignoring the law of reciprocal actions). This aspect of the answers is generally observed for the "states" version (box 2).

In other cases, F_{ext}.(the "external" force) is linked to the displacement of the lower end of the spring as if the extension concerned only the lower spring: $F_{ext}=k_1\Delta l$. This typical response is often associated with a diagram of forces that is compatible with the correct analysis: all the force magnitudes are equal, which was not the case previously, when the external force alone had to compensate for two resistances. Here the image is that of a *transmission* (dare one say "propagation"?) of forces, rather than of a static opposition. This characteristic is found mostly in answers for the "transformation" version (see box 2).

Box 3

Common answers to the questions in box 2

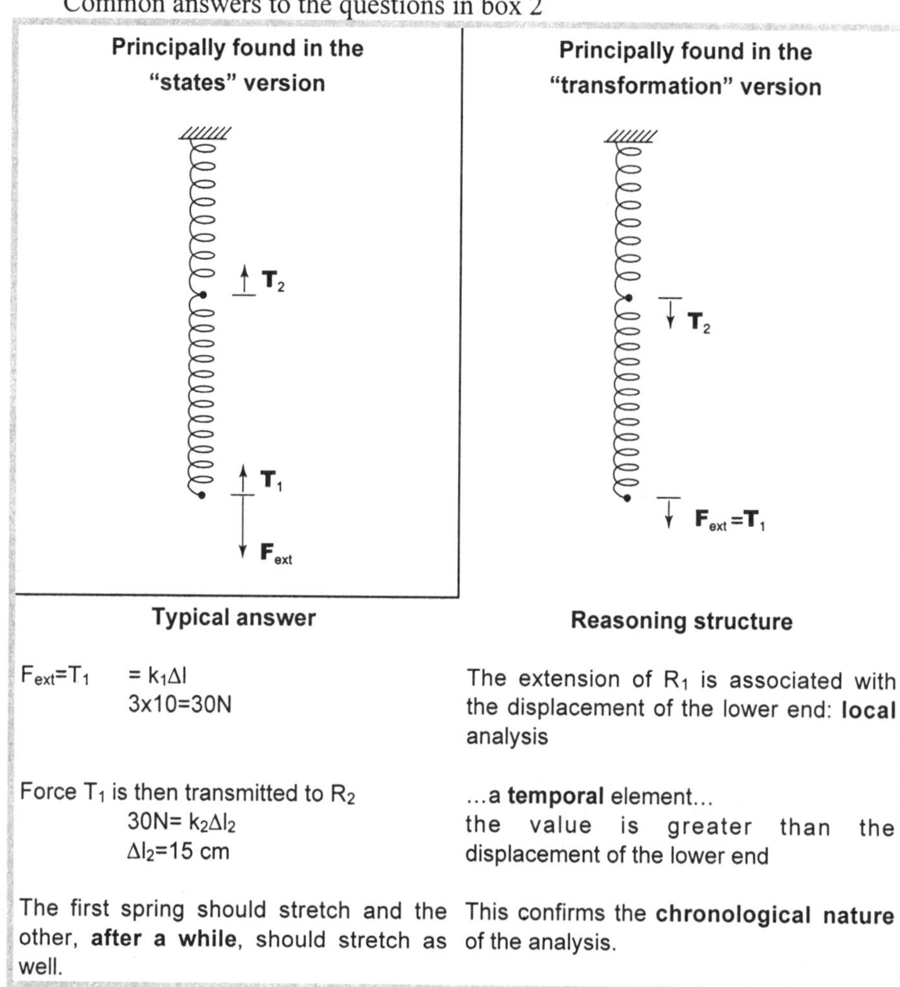

Typical answer	Reasoning structure
$F_{ext}=T_1 \quad = k_1\Delta l$ $3x10=30N$	The extension of R_1 is associated with the displacement of the lower end: **local** analysis
Force T_1 is then transmitted to R_2 $30N= k_2\Delta l_2$ $\Delta l_2=15$ cm	...a **temporal** element... the value is greater than the displacement of the lower end
The first spring should stretch and the other, **after a while**, should stretch as well.	This confirms the **chronological nature** of the analysis.

Taking this analysis further, the author points out that, in the erroneous relationship $F_{ext}=k_1\Delta l$, the fact that the junction point of the two springs is also displaced by a downward pull is not taken into account. Writing $F_{ext}=k_1\Delta l$ instead of $F_{ext}=k_1\Delta l_1$ probably indicates a very local analysis of events, focusing only on the spot where the external action is applied. Only later is the transmission of the action to the rest of the system envisaged. This gives curious results, as overestimating the lower spring's extension entails overestimating the intensity of the forces as well, and the extension of the upper spring turns out to be greater than the displacement of the lower end (see the student's answer included in box 3).

But what is most striking is the role of time in this type of solution. The commentary in box 3 clearly shows that the external force is seen as having first a local action, and then, "after a while, it is transmitted to the top spring". This is a clear denial of the premises of quasistatic analysis. It is hardly surprising that this sort of denial occurs most often in a transformation situation.

The recurrent nature of these elements in students' answers makes it clear that the spatio-temporal content of the situations and questions is important in natural reasoning. This is confirmed through other interesting examples in the same study.

A physical situation involving communicating vessels is proposed (Fauconnet, 1981) in questions comparable to those we have just described; again, there are two versions of the questions. For this situation, the results are surprisingly similar as regards both the typical elements found in the answers and the frequency with which they arise in the "states" and "transformation" versions.[1]

As we shall now see, investigations in another, much broader field have led us to conclusions which complement remarkably this pioneering study.

3.2 Sequential reasoning in electric circuits

Electric circuits are structures that have a clear spatial content. Box 4 shows a series of situations that all have a common structure. In each situation, two identical components (elements of a circuit with terminals) frame another, different circuit element, and a generator. The equations that

[1] The proposed situation is as follows: a liquid is poured into two vessels of different cross-sections; the bottoms of the vessels communicate. A quantity of non-miscible liquid of volume ΔV is added in one vessel. The volume ΔV of the "external" transformation agent is often associated with the cylindrical volume between the initial and final positions of the separating surface. In other words, it is seen as indicating a *local* transformation; the displacement in the level of the liquid on the other side of the system is not taken into consideration.

apply to series circuits like these in quasistationary analysis are contained in box 1. They clearly show that the order of the elements is irrelevant. If one dipole is exchanged for another, the values of the affected quantities do not change; neither are the other components affected in any way.

Box 4

Sequential reasoning at work (from Closset, 1983)

Situation	Example of a sequential answer	Proportion of sequential answers[a] E1	E2	E3
Situation 1 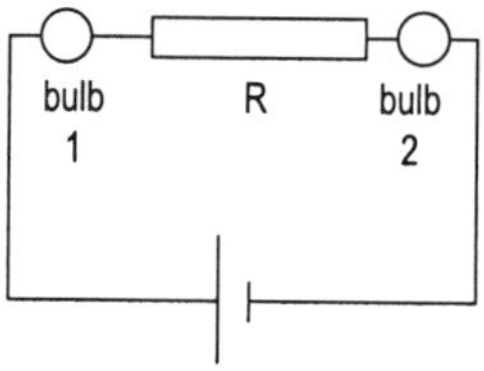 (bulb 1, R, bulb 2)	The second bulb shines less brightly	51%	52%	10%
	N =	91	52	96
Situation 2 (C_1, R, C_2) 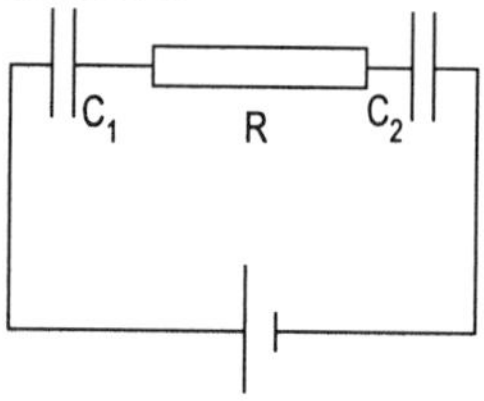	The second capacitor is charged more slowly	68%	33%	37%
	N =	55	82	109
Situation 3 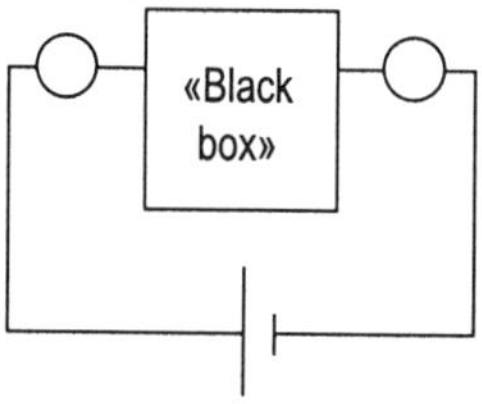 («Black box»)	For the two bulbs to shine equally, there must be no battery in the box	52%	46%	33%
	N =	58	83	102
Situation 4 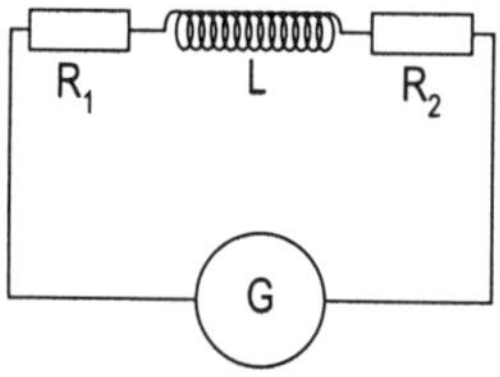 (R_1, L, R_2, G)	The potential difference across R_2 is out of phase with the potential difference across R_1	50%	38%	22%
	N=	18	16	81

Situation	Statement	E1	E2	E3
Situation 5 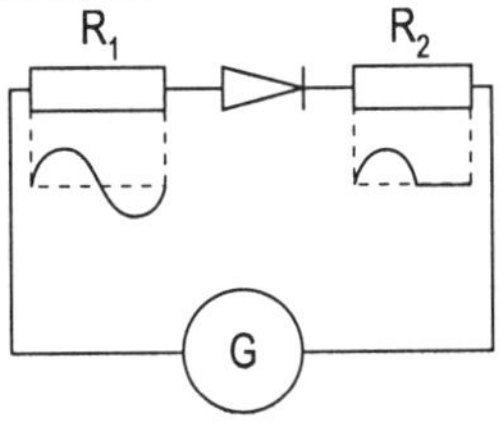	The potential difference across R_2 is redressed, and not the potential difference across R_1	28%	41%	24%
	N =	18	16	81
Situation 6 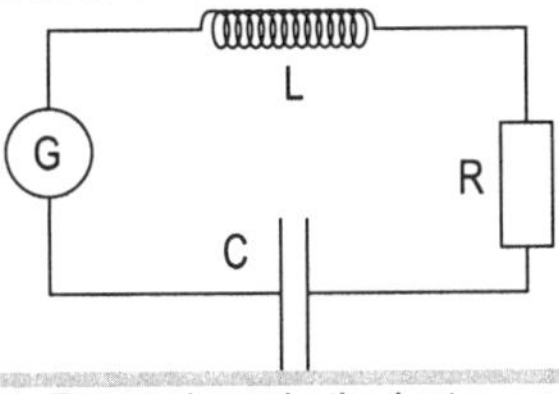	The phase of the potential difference across the three dipoles depends on the order in which they are placed		24%	17%
	N =		138	89

a. E1: students in the last year of secondary school in Belgium; E2: first year university students; E3: students with two to four years of university education.

The comments of the pupils and students very often deviate from this analysis. In the first situation, for instance, they often say that the second bulb shines less brightly than the first, which is, nevertheless, identical. In the next situation, they state that the second capacitor is charged less rapidly than the first after the circuit has been closed (from the first billionth of a second, and even earlier, quasistationary analysis is amply justified), although the two capacitors are charged at the same rate.[2] In the question concerning the black box, they are asked if bulb 1 shines in the same way as bulb 2 no matter what is in the box (which is indeed the case), or if it depends on the contents of the box. Several conditions are stated: "No battery in the box," "No resistor..." When there is a resistor at either end of a coil, the difference in phase of the potential difference across the resistors relative to the potential difference across the generator is said to be different: "There is a change of phase for the second resistor, the potential difference before the coil is not affected." And in the situations on alternating current, the circuit in which two resistances frame a diode often gives rise to the idea, shown in the diagrams, that the potential difference is rectified after the diode and not before it, whereas the diode affects the current in the circuit as a whole; and therefore also affects the potential difference across all the resistances in series.

All these answers are described as "sequential" in the box, because they can be interpreted as stemming from the following reasoning:[3]

[2] This is true even when the capacitors are not identical.

[3] See also Shipstone (1985).

There is an entity referred to somewhat interchangeably as "electricity" or "electrons," to which students associate various physical quantities, such as current, potential difference (or voltage), or even phase. It leaves the generator through one terminal and heads out into the circuit. Its progress is more or less affected as it passes each component; some students explain, for example, that "the current gets used up in the resistor," or that "the current is rectified by the diode", but there is no action "backwards" from the end point to the source. Generally, the entity returns to another end of the generator, though the adventures described do not always end that way.

There are variations on this theme. Sometimes the local character of this type of reasoning is so strong that each component and the quantities associated with it are considered independently from the other components. Another particularly tenacious idea is that the current leaving the generator is the same no matter what the circuit, like a spring that always yields the same amount of water regardless of whatever dams there may be further down the river.[4]

That such an idea should outlive higher education casts doubt on the simplistic interpretations of Bachelard's "epistemological break"[5] – an idea echoed (perhaps rather dubiously) by the recent interest in "conceptual change" in the Anglophone literature (see Posner et al., 1982). The transition from common knowledge to scientific knowledge is held to involve crises, cognitive conflicts that bring about decisive or even definitive restructuring. The survey results presented above scarcely bear out the theory that such cognitive overhauls occur.

Of course, advanced students no longer believe that two identical bulbs in series do not shine equally brightly; they have mastered that highly familiar situation. But, faced with an unfamiliar question, such as the one concerning the "black box" (box 4, situation 3), a great many of them rejoin the ranks of sequential reasoners. As J.L. Closset (1983) put it, "Sequential reasoning does not disappear, it is suppressed." For these advanced students, then, there has been no dramatic restructuring, but, rather, local learning. Even teachers sometimes show that sequential reasoning is their most readily available cognitive tool. Such reasoning certainly does work quite often. For proof that a systemic vision of a circuit is not a requisite for everyone, one has only to listen to electronic engineers talking about frequencies that "start", "pass" or "do not pass," without ever mentioning that anything

[4] All these difficulties are now explicitly addressed in France. On this subject, see the French National Curriculum for grade 8 (*Bulletin Officiel*, 1992a, applied in 1993) – see appendix to chapter 2 and Couchouron et al. (1996). For a compilation of consciousness-raising questions, see Courdille (1991).

[5] A notion which resembles that of the "change in paradigm" that Kuhn introduced in epistemology.

"comes back" through the electrical Earth or that the circuit is closed. The comment of Bernard Schiele seems to fit the data: "The difference between scientific knowledge and common knowledge is a matter of degree, not of kind" (Schiele, 1984, p 91; see also Viennot 1989b).

These two examples of systems with a clear spatial structure indicate that reasoning is influenced by the way one looks at things. With electric circuits, in particular, time is introduced naturally in the typical analysis of the proposed situations. In such an analysis, students follow the progress of the main character, "electricity," around the circuit suggested by the diagram. It must be pointed out that in this case, natural reasoning is "based", so to speak, on an artificial support, i.e., the way circuits are represented in diagrams at school. Children who have never seen such diagrams do not make the same mistakes. When they have never seen a closed circuit, they imagine other routes, or reason along totally different lines, asserting, for instance, that two identical bulbs will shine equally brightly, because "they take what they need".[6]

3.3 Heat conduction

From Rozier (1988)

Other situations also show how spatial structure can determine reasoning. The conduction of heat along bars, for example, often gives rise to sequential reasoning. In the survey question presented in box 5 (Rozier, 1988), regarding the transfer of heat along a conductor made up of two sections with different conductivities, students are asked about the consequences of replacing the second (or "downstream") part.

As might have been expected, the most common answer is that only the quantities concerning the "downstream" part change. Yet the system of equations applying to this problem (a quasistationary situation) indicates clearly enough that the quantities in the two parts cannot be calculated separately: the "downstream" part affects the "upstream" part. But, once again, "conduction" brings to mind "forward movement." This time, the leading actor in the story is "heat," seen as an almost local entity whose evolution in space takes time and differentiates the downstream from the upstream level.

[6] See Closset, 1983, pp 202-203.

Box 5

The forward movement of heat (from Rozier, 1988)

There is a flow of heat along a bar of thermal conductivity λ, in the **x** direction (see diagram).

The temperatures of the ends of the bar are fixed and equal to T_1 and T_2 ($T_1>T_2$).

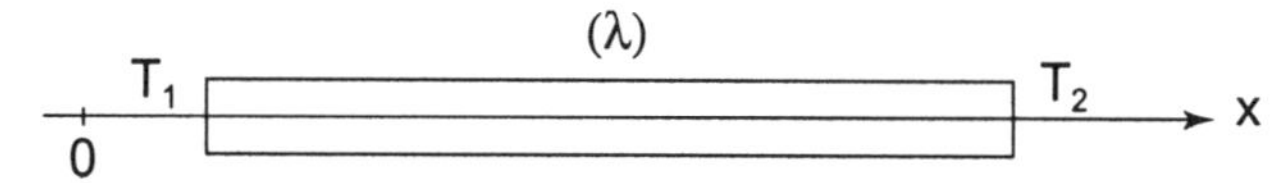

One part of the bar is replaced by another bar, of the same cross-section, but with thermal conductivity $\lambda' \neq \lambda$ (see diagram).

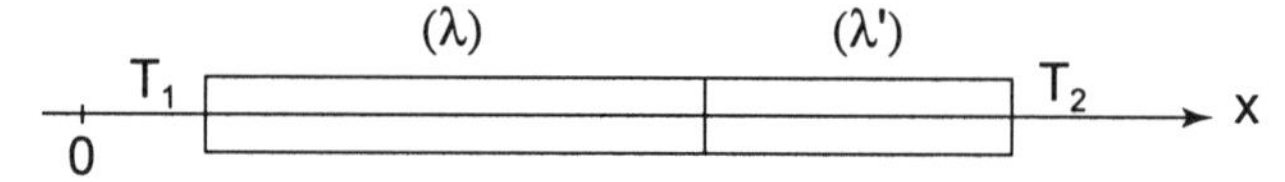

Do the temperatures in the part of the bar that has not been changed remain the same? (The bar is in a steady state)

- ❑ Yes
- ❑ No
- ❑ I do not know

Results (N=106[a])

Incorrect, "sequential" answer (YES)	35%
Correct answer (NO)	54%
No answer	11%

Summary of correct answer[b]:

1st case: homogeneous bar, steady state — 2nd case: two-part bar, steady state

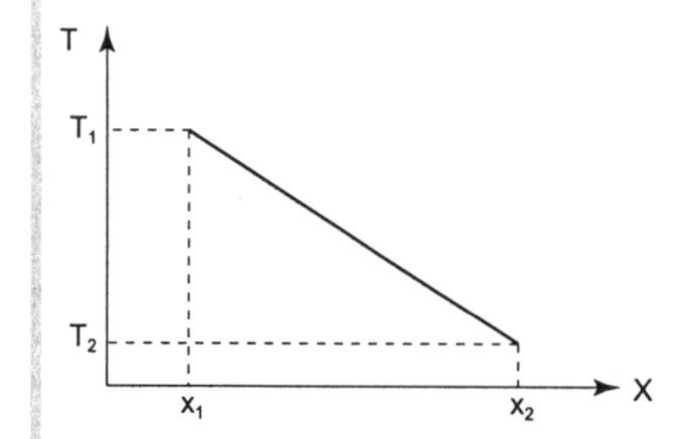

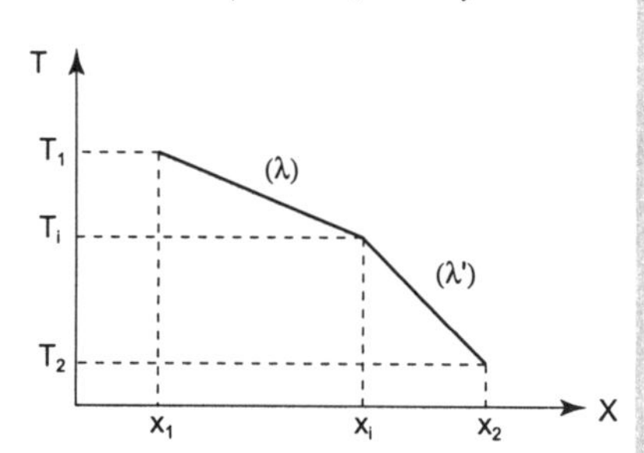

Typical comments

- "Flow of heat from 0 towards x: the second part has no effect on the first part";
- "The heat flows through the unchanged portion of the bar normally, up to the point where change begins and where the temperatures of the different points are changed."

a. Students preparing for "Grandes Ecoles" (France): Mathématiques Supérieures (N=62) and Mathématiques Spéciales (N=44), combined results.

b. Temperature T_i is determined by the fact that the flow of heat must be equal on both sides of the junction (x_i).

4. SYSTEMS WITH NO CLEAR SPATIAL STRUCTURE: EXAMPLES FROM THERMODYNAMICS

From Rozier (1988); see also Rozier and Viennot (1991)

What happens when no particular "upstream" or "downstream" direction is suggested? Does time still come into spontaneous reasoning? If so, what is the leading actor whose adventures are imagined?

In thermodynamics, quasistatic analyses are commonly applied to multi-variable systems. Asked whether they know any multi-variable problems in physics, the great majority of first and second year university students bring up thermodynamics straight away. It is as if they had not noticed that practically all problems in physics involve multiple variables.

4.1 Linear reasoning

Box 6 summarises a survey question on a classical situation: the compression of a perfect gas, with no exchange of heat with the environment ("adiabatic" change), carried out in a quasistatic manner. As indicated, pressure and temperature increase. The students are asked to explain these phenomena "in terms of particles," that is, by referring to the properties of the atoms or molecules that constitute the gas. The correct answer is outlined in box 6, simply to indicate to the lay reader that several quantities are involved at each stage of the argument. The volume decreases, and therefore the concentration of particles increases; moreover, the action of the piston increases their speed (more precisely, their average speed). These two factors contribute to an increase in the number of collisions with the sides of the container (per unit area and per unit time). This, together with the "force" of the impacts (which is associated with the linear momentum of particles, and thus with their speed), explains the increase in pressure.

A different form of reasoning is apparent in many of the answers (42%, N=111). They can be summed up as follows:

The volume decreases→the number of particles per unit volume increases→the number of collisions increases→the pressure increases.

At each stage, only one quantity is considered, and the associations that are established are binary: a (single) quantity determines another (single) quantity. In this way, a preference for certain factors is established. Thus, pressure is generally associated with a geometrical factor, the concentration of particles (or "squashed" particles), and the dynamic aspect is forgotten.

"Preferential associations" of this sort are also evident in the answers to many other questions.

Box 6

Example of causal linear reasoning applied to the adiabatic compression of a perfect gas – From Rozier (1988).

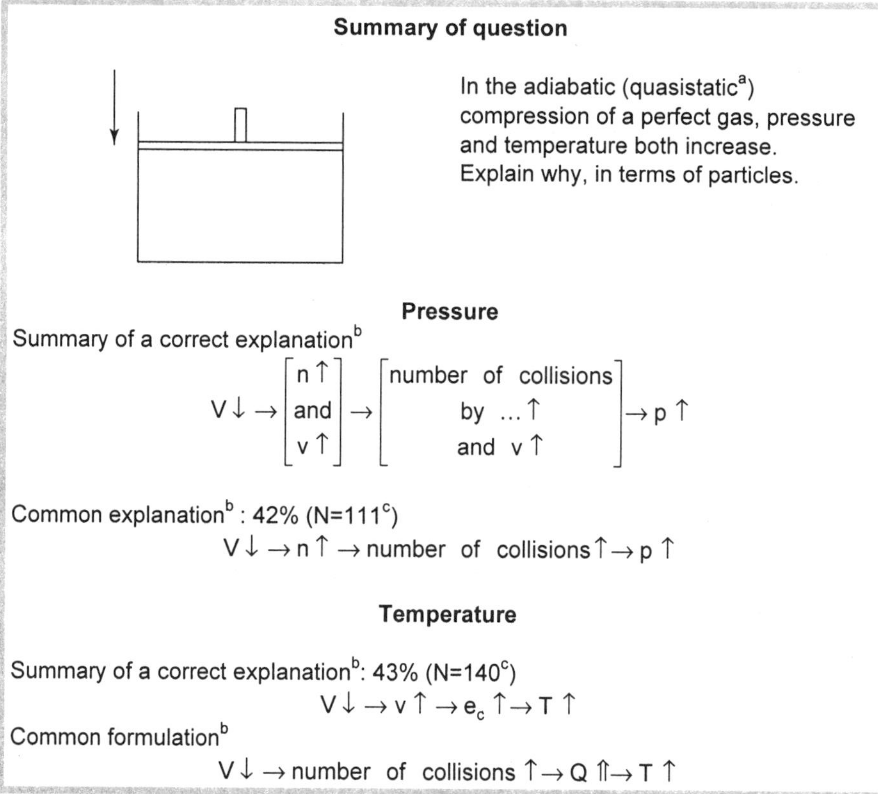

a. The results are the same whether or not the quasistatic nature of the compression is stated in the question.
b. Key: Volume of gas: V; number of particles per unit volume: n; pressure of gas: p; temperature of gas: T; average speed of the particles: v; mean kinetic energy of a particle: e_c; heat: Q; "increases" ↑; "decreases" ↓; "produced" ⇑ ; "therefore":→.
c. Sample: first to third year post-baccalaureate (the high school leaving diploma) students, treated as a single group.

Several other studies confirm these results[7] and justify the hypothesis that common reasoning frequently adopts the structure of a linear chain of implications, in which each link is a single phenomenon (ϕ), bearing on the evolution of a single quantity:

$$\phi_1 \rightarrow \phi_2 \rightarrow \phi_3 \rightarrow \ldots \phi_n \rightarrow \phi_{n+1} \rightarrow$$

[7] Rozier (1986), Maurines (1986), Maurines (1991).

This linearity in reasoning is massively apparent. How such arguments are constructed has been discussed in the preceding chapters (chapters 3 and 4): to explain a particular motion, there has to be a cause, but one cause is enough.[8]

Once again, we can observe the natural tendency of reasoning to simplify analysis by reducing the number of factors to be considered.

The example just given illustrates a simple "oversight" (more precisely, one aspect of the problem has not been taken into account: the "force of the collisions").[9]

It sometimes happens that each single quantity or phenomenon that makes up a link in the chain owes its "simplicity" to the fact that it is a compound, an undifferentiated notion, which combines several concepts from physics. For example, force, velocity, energy, and so on are fused as "supply of force" (Viennot, 1979); the height and the velocity of a pulse on a rope in a "supply of stored signal" (Maurines, 1986); current, potential difference and energy as "electricity" (Johsua, 1985; Johsua and Dupin, 1989; Closset, 1983 and 1989); heat and temperature as "heat" (Tiberghien, 1984c; Rozier, 1988); speed and distance between particles as "thermal motion" in a gas (Rozier, 1988).

In other situations, linear reasoning may be applied to an entire set of relevant quantities, which are then analysed in a step-by-step, variable-by-variable manner.

4.2 Time and causality

A questionnaire from the same survey takes things one step further. The question summarised in box 7 presents a classical situation in elementary dynamics: the heating of a gas at constant pressure. Once again, the correct answer is discouragingly complex. The most common answer, also summed up in box 7, is surprisingly common (43%, N=120), and has the by now familiar linear structure:

Heat is supplied→the temperature rises→the pressure increases→the volume increases.

But what is truly surprising here is the equanimity with which the students contradict data presented in the problem, namely, that the pressure is constant throughout.

Some explicit comments provide a key to this enigma. Taken together, they give the following explanation:

[8] See also Ogborn (1992), Andersson (1986), Driver et al. (1985).

[9] See also Méheut (1994).

"First of all, the piston is blocked; heat makes the temperature and pressure rise; then the piston is free to slide, the volume increases, and the pressure returns to its external value."

Box 7

Causal linear reasoning and contradicton of given data – From Rozier (1988)

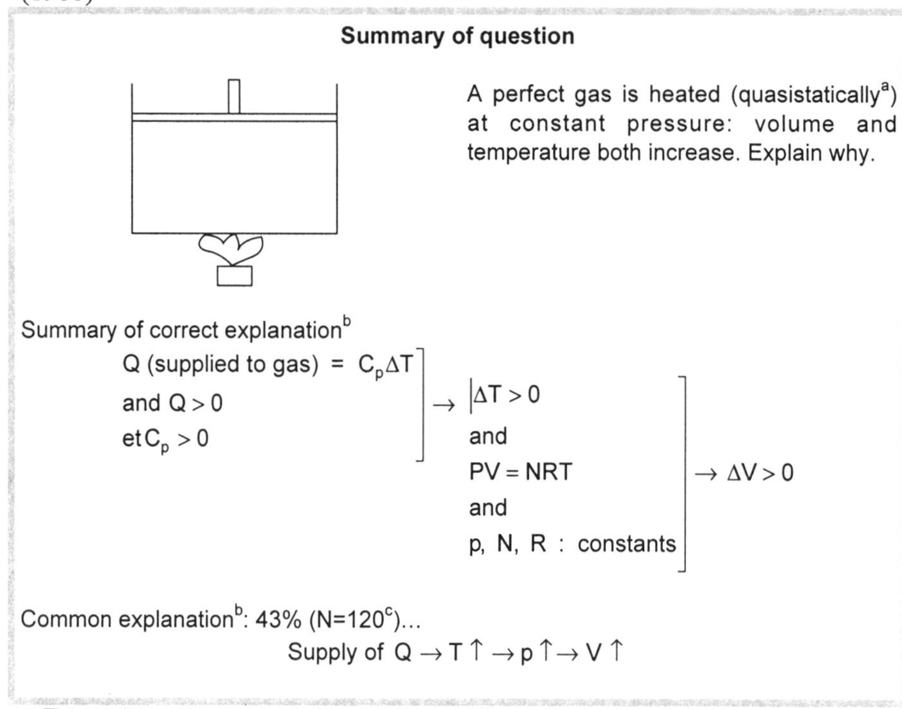

a. The results are the same whether or not quasistatic heating is stated in the question.
b. Key: see box 6; c_p: molar specific heat capacity at constant pressure; R: perfect gas constant; N: number of moles.
c. Sample: first to third year post-baccalaureate, grouped.

In other words, the argument is based on a temporal sequence. The arrows in the explanatory summary proposed above do not only indicate a logical link. They have a chronological connotation, which is made more or less explicit.

It sometimes happens that the single quantity or phenomenon – each link in the chain – owes its "simplicity" to the fact that it is a compound, an undifferentiated notion, which combines several concepts from physics. For example, force, velocity, energy, and so on are fused in "a supply of force" (Viennot, 1979); the height and the velocity of a wave on a rope in a "supply of stored signal" (Maurines, 1986); intensity, tension and energy in an "electric current" (Johsua, 1985; Johsua and Dupin, 1989; Closset, 1983 and

1989); heat and temperature in "heat" (Tiberghien, 1984c; Rozier, 1988); distance and speed between particles in "thermal motion" in a gas (Rozier, 1988).

Moreover, linear reasoning is sometimes applied to an entire set of relevant quantities, that are then analysed in a step-by-step, variable-by-variable process.

In English, French, Spanish, and probably many other languages as well, there is an intermediate term between the expression of a logical link ("therefore") and of a temporal succession ("later"). In English, the logical and chronological levels meet in a single word, "then". And to what extent do we ourselves know, when we say "then," or "alors," what we mean by that adverb?

Table 1
Different languages, same ambiguity in meaning

Level	French	English	Spanish
Logical	donc	therefore	por eso
Intermediate	alors	then	entonces
Chronological	ensuite	later on	despues

Similarly, who knows to what extent the future tense is to be taken literally in a statement such as: "The pressure increases, the piston will slide"? Isn't the future tense commonly used in teaching to describe what takes place, when it is presumably a consequence of something else ("If I reduce this quantity, this other one will increase," etc.)?

Common explanations are steeped in time. Even in the absence of a suggestive spatial structure, we often tell stories about systems. When a complex object is transformed as a whole, we analyse the phases one by one, as if each one acted upon the other, and took some time.

4.3 Difficulties in the analysis of steady states

Once again, the point which follows is based on a survey question. It concerns the second part of the questionnaire described in box 6, on the adiabatic compression of a perfect gas. Why does the temperature rise? Common explanations of this phenomenon (43% of the students interviewed, N=140) can be schematised as follows:

The volume decreases→the number of particles per unit volume increases→the number of collisions increases→heat is produced→the temperature rises

There is an explicit link between the number of collisions and the production of heat in approximately half of the cases: "The collisions between molecules produce heat." This assertion deviates from accepted physics. Indeed, if it were the case, energy would keep on accumulating in the container, as heat cannot escape from it (it is "adiabatic"). This could not go on for very long.

But duration is precisely what those who consider collisions as generating heat are least concerned about.

Because they are seen as *successive*, the events described in the explanation are considered as *temporary*. The story of the collisions does not last long in the argument; nor do the students imagine it to last long when considering the phenomenon. Permanence is not envisaged, nor, consequently, is the fact that such a phenomenon would diverge from what is reasonable.

It is surprising that when teachers in training sessions are asked how to make a student who believes in this economical heating system see that he/she is mistaken, they hardly ever think of this divergence as a counter-argument (Viennot, 1993). Imagining a steady state does not come naturally to them either. And when that objection is presented to students, they frequently reply: "The collisions do not release any more heat than during the compression." They might as well say, "Come to think of it," or "The only thing I am interested in is the extra heating. That is enough to explain the increase in temperature." But introducing the idea of divergence is, in the end, the most effective argument, because it addresses one of the main axes of common reasoning: the division of stories into episodes.

4.4 Steady states of non equilibrium: the example of the greenhouse

Predictably, the difficulty in taking duration into account leaves its imprint on common explanations of steady states of non-equilibrium. These are phenomena which involve constant flows of energy into and out of a system. The state remains constant when the flows compensate one another. Otherwise, the state is transient and the system under study gains or loses energy globally. The greenhouse effect is an illustration of this.

It is striking how often the explanations of this effect found in various popularisations describe a transient state without specifying it: "It is hotter in a greenhouse because more energy enters it than leaves it (because of the properties of the glass)". Here again, we might ask how long this state of affairs could last without endangering the gardener's health. But this question is not natural; it seems that as soon as one explanation has been hit upon, the case is closed. That is that. End of question, never mind duration.

Some trainee teachers were asked to draw diagrams explaining the greenhouse effect (N=85). Box 8 gives a few representative answers.

Box 8

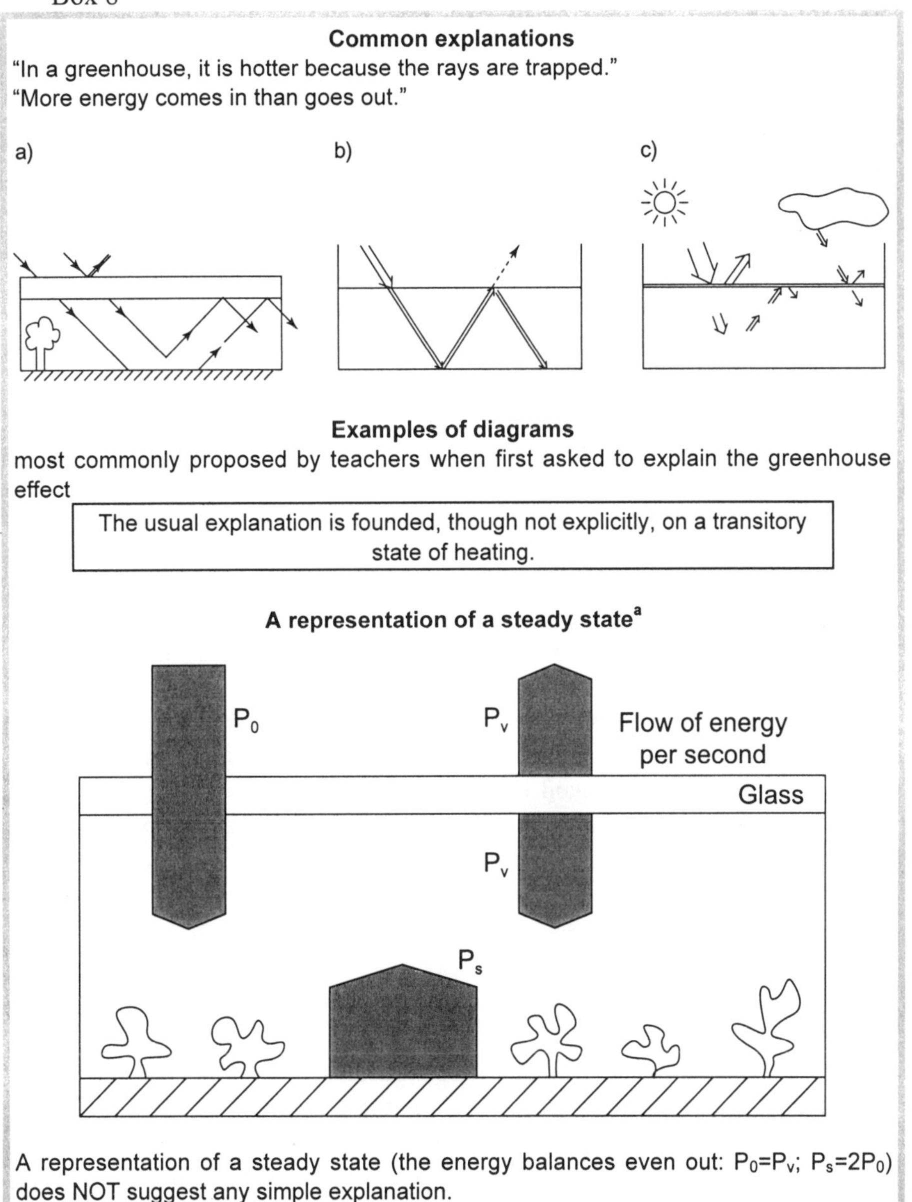

a. This representation is coherent but overly simplified: it corresponds to a situation in which the incident rays pass completely through the glass, while the ground emits upward beams that are totally absorbed and re-emitted by the glass (which acts like a "black body").

In hardly any of these answers is the incoming energy described as equal to the outgoing energy. What these diagrams do show is the arrival of incident rays; they go through a series of adventures: one fraction is reflected, another is transmitted towards the ground, then the ground sends a ray towards the window pane and a part of this ray heads back towards the ground, and then..., etc. Some speak of "trapped rays." The story contains a great many events that are correctly analysed. But there is no overview; no one considers the flows entering and leaving the system simultaneously. And no one finds this explosive greenhouse disturbing in the least.

Those who give explanations of this kind are always surprised when they realise that it can be hotter inside a greenhouse than outside it, even when less energy enters it than leaves it, as happens at night, for example.

But in any attempt at explaining the greenhouse effect verbally, difficulties arise. What comes naturally, even in a correct explanation, is a description which deals with energy balances and variations in temperature successively, as in the following statement:

"The system, glass panes included, receives a little more energy than it can emit, because of the transmission band of the glass; therefore, its temperature rises, so that it emits a little more energy in this wave band; and so the balance is still not even, but closer than it was, etc., until the same amount of energy gets out as comes in," with the same sort of argument to explain how the greenhouse cools.

Physicists use the differential method to legitimise the approach suggested in the explanation above, and make it work for them. Taken to the limit, this procedure leads to analytical results that observe the constraints summed up earlier: permanent relationships, simultaneous evolutions. That is the price that has to be paid if the initial desire for an explanation (more energy comes in than goes out) is to be replaced by a description of a permanent state, which does not explain anything at all. It is, conceptually speaking, a very high price. Understanding such a procedure is no simple matter; it is even difficult to see what it achieves (Artigue et al., 1990, 1991).

4.5 Linear causal reasoning and understanding texts[10]

As regards this example of the greenhouse, one could lay the blame on the offhand nature or carelessness of common explanations, and leave it at that. But then one principal factor would be ignored: the satisfaction that people get from such explanations. And when a text contains no error in physics, but is misunderstood because of a proximity that "echoes" the linear causal reasoning of the students, the appeal of this type of reasoning is particularly apparent.

[10] Excerpt from an article published in *Didaskalia* (Viennot, 1993), based on a study by Rozier (1988).

Students in higher education (Math. Sup, Math. Spé, Licence and I.U.T.[11] students taken as a group) were asked to read the following text (Valentin, 1983):

> The mean energy of motion of each molecule is enough to prevent the molecules of gas that surround us from being bound to each other: in a gas, molecules are continually colliding and rebounding. But, if the temperature is lowered, the system will be able to become liquid and even solid. These physical phenomena occur when, with decreasing temperature, the kinetic energy of the molecules becomes so low that they can no longer resist the electromagnetic interaction. They first gather in the liquid state and finally get bound in the solid state.

The students were then given the following passage to read:

> At any given moment during liquefaction, the mean kinetic energy of a molecule of gas is larger than the mean kinetic energy of a molecule of liquid (liquid and vapour are in thermal equilibrium at the time considered).

The students are asked if they feel that the text suggests such a statement, and if the statement is true. An overwhelming majority of students consider that the text does suggest such a statement (77%, N=181), and that the statement is true (80%). Yet on the next page, the author makes the point that in thermodynamic equilibrium, the mean kinetic energy of molecules is the same in the liquid and gas phases. The statement the students were given to assess is, therefore, false, and was not meant to be conveyed in the text. And yet the students thought that was what they had read. How is one to interpret this result?

Cautiously, for a start. We are working from conjecture and our conclusions cannot be final. For example, it might simply be that, as the students think the statement is true, they believed they had read what they already thought to be true. But why do they think so?

On closer examination, one sees that the passage is punctuated with temporal connotations: "If... will be able to... when... so low that... they can no longer... first... finally...". This structure can be summed up in the form of a logical chain:

(gas) T falls→e_c (mean kinetic energy) decreases→the electromagnetic attractions win→liquid state→ ... →solid state.

[11] In France, Mathématiques Supérieures and Mathématiques Spéciales are respectively the first and second years of preparation for the competitive examinations for admission to institutions (*grandes écoles*) specialising in science. I.U.T.: Institut Universitaire Technologique

The common tendency to insert time into explanations is encouraged here. The text can be read as a story. First, there is a gas; then, there is a liquid. In between, the temperature, and then the mean kinetic energy of the molecules, diminishes. In the statement, the two phases are in equilibrium, i.e., are present simultaneously, at the same temperature. It would seem, from the students' answers, that the chronological structure they see in the text has overshadowed the idea of simultaneity that the excerpt intended to convey; it apparently leads to the belief that the mean kinetic energy is lower in one phase (the "final" phase) than in the other (the "first" phase).

The complexity of this analysis may elicit a certain scepticism. But one cannot remain indifferent in the face of the massive rate of erroneous readings. These results are a reminder that, although natural reasoning often goes unnoticed, it is not without impact.

5. LINEAR CAUSAL REASONING AND QUASISTATIC APPROACHES: IRREMEDIABLE DIFFERENCES?

The forms of common reasoning identified by Sylvie Rozier can be summed up as follows: the common tendency is

- to use linear arguments, linking phenomena, each of which relates to the evolution of a single physical quantity:
  $\phi_1 \rightarrow \phi_2 \rightarrow \phi_3 \rightarrow \ldots \phi_n \rightarrow \phi_{n+1} \rightarrow$
- to superimpose, more or less explicitly, an effective chronology onto the argument, thereby giving it the status of a story.

This type of reasoning can apply to systems with or without a clear spatial structure.

If the spatial structure of the system suggests a journey, the "hero" of the story is often an entity in which several quantities are vaguely associated, or "stuck together." The complexity of the analysis is then played out in space or time: the entity has various adventures, linked like the episodes of a story.

If the system is a complex object undergoing simultaneous changes, the transformations of the separate single quantities that characterise it are considered separately. The explanation is given the status of a sequence of events.

These events are seen as both successive and temporary. Hence the difficulty in taking into account non-equilibrium steady states, where the incoming and outgoing flows are always equal.

Sylvie Rozier makes the interesting hypothesis that, in natural reasoning, the complexity of the spatial analysis and the complexity of the analysis of variables interfere with each other, i.e., both cannot be fully developed. At this level of conjecture, it is difficult to provide data likely to convince everyone. But it is true that all the results included in this book support her theory.

No specific teaching strategies emerge from these results. Mastering of multifunctional reasoning is a necessary goal, since all of life consists of multi-variable problems. But how to begin, and when? This is typical of the kind of general theme that is not linked to any one physics topic but applies to many. It can only be the object of a long-term commitment, and of instruction covering various aspects of physics (see also chapter 9).

But it can be introduced much earlier than one might imagine, from the relatively sophisticated examples mentioned above. To form a mental picture of something as simple as a rectangle of rubber being stretched, two variables have to be considered, length and width; does the width remain constant while the length increases? Such questions can be put to pupils as soon as they have learned the "formula" for the area of a rectangle. Recognising that unless something is known about the width of the rectangle, nothing can be said about how the surface evolves, is the first step towards a much-needed modesty: the beginning of wisdom, as far as multi-variable problems are concerned, it to be able to detect cases where nothing can be asserted or explained.

An awareness of the complexity of such problems may encourage two teaching strategies.

The first is the one we have just described: to specify what cannot be accounted for. This is, no doubt, the right thing to do when dealing with changes of state. They can be described, but not explained in terms that speak to the imagination: Why is it that in thermodynamic equilibrium, the molecules of a liquid are "bound," whereas gas molecules are "free," when they all have the same mean kinetic energy per particle?[12]

Admitting that there are limits to our powers of explanation might be a good thing in other areas, such as economics, too.

The second strategy is to establish clear boundaries for partial explanations. One can say, for example, that "at high altitude, pressure is lower because there are fewer molecules," if one specifies that this type of explanation is valid only when there is no (significant) variation in temperature, or that the lower density of particles is the principal factor here, but that there can be other important factors in other situations, etc. The expression "all other things being equal", which is so dear to physicists, has to be explained when it is a condition upon which a conclusion depends, but the conditions in which this applies must be recognised as limited.

[12] This is true in classical description, that is to say in the conditions where the theorem of equipartition of energy applies: then, the mean kinetic energy per particle depends only on temperature for a given type of particle.

These attitudes, if adopted very early on in teaching, at least prepare the student for multi-variable problems, and discourage inconsistency.

In France, they have been explicitly recommended as teaching aims, to be introduced as early as grade 9 (Physics GTD, 1994). Indeed, at that stage, the relationships students need to master completely, such as Ohm's law or the laws of electrical power, all involve more than two quantities.

In fact, the question is how to make these relationships express what they mean, no more, no less. This is demanding and interesting. Demanding, because the overly reductionist formulations of single-variables are abandoned. Interesting, because one learns that, in physics, one can establish non-trivial relationships and reconcile what first seemed to be contradictory assertions by using more precise formulations.

"Qualitative" is definitely not synonymous with "sloppy", especially when, as in the difficult area of multifunctional relationships, the objective is to distinguish between pseudo-demonstrations and conclusions that are based on a rigorous study of physical theory.

Nor should taking such a "hard line" when applying qualitative reasoning necessarily be delayed until students have nearly completed their scientific training.

REFERENCES

Andersson, B. 1986. The experiential Gelstalt of Causation: a common core to pupils' preconceptions in science. *European Journal of Science Education*, 8 (3), pp 155-171.

Artigue, M., Ménigaux, J. and Viennot, L. 1990. Some aspects of students' conceptions and difficulties about differentials. *European Journal of Physics*, 11, pp 262-267.

Artigue, M., Ménigaux, J. and Viennot, L. 1991. *Questionnaires de travail sur les différentielles*. Université Paris 7 (Diffusion IREM et LDPES).

Bulletin Officiel du Ministère de l'Education Nationale 1992a, n°31, Classes de quatrième et quatrième technologique, pp 2086-2112.

Closset, J.-L. 1983. *Le raisonnement séquentiel en électrocinétique*. Paris, Thèse, Université Paris 7.

Closset, J.-L. 1989. Les obstacles à l'apprentissage de l'électrocinétique. *Bulletin de l'Union des Physiciens*, 716, pp. 931-950.

Couchouron, M., Viennot, L. and Courdille, J.M. 1996. Les habitudes des enseignants et les intentions didactiques des nouveaux programmes d'électricité de Quatrième, *Didaskalia,* n°8, pp 83-99.

Courdille, J.M. 1991. *Questionnaires de travail sur l'électrocinétique* Université Paris 7 (Diffusion IREM et LDPES).

Driver, R., Guesne, E.and Tiberghien, A. 1985. Some features of Children's Ideas and their Implications for Teaching, in Driver, R., Guesne, E. and Tiberghien, A. (eds): *Children's Ideas in Science*. Open University Press, Milton Keynes, pp 193-201.

Fauconnet, S. 1981. *Etude de résolution de problèmes: quelques problèmes de même structure en physique*, Thesis (Thèse de troisième cycle), Université Paris 7.

Gutierrez, R. and Ogborn, J. 1992. A causal framework for analysing alternative conceptions, *International Journal of Science Education*. 14 (2), pp 201-220.

Johsua, S. 1985. *Contribution à la délimitation du contraint et du possible dans l'enseignement de la physique (essai de didactique expérimentale),* Thesis (Thèse d'état), Marseille, Université de Provence.

Johsua, S. and Dupin, J.J. 1989. *Représentations et modélisations: le "débat scientifique" dans la classe et l'apprentissage de la physique*, Peter Lang, Berne.

Kuhn, T.S. 1973. *La structure des révolutions scientifiques*. Paris, Flammarion.

Maurines, L. 1986. *Premières notions sur la propagation des signaux mécaniques: étude des difficultés des étudiants*. Thesis. Université Paris 7.

Maurines, L. 1991. Raisonnement spontané sur la propagation des signaux: aspect fonctionnel. *Bulletin de l'Union des Physiciens*, 733, pp 669-677.

Méheut, M. 1994. Construction d'un modèle cinétique de gaz par les élèves de collège. Jeux de questionnement et de simulation. *Actes du Quatrième Séminaire National de la Didactique des Sciences Physiques*, IUFM d'Amiens, Amiens, pp 53-73.

Posner, G., Strike K., Hewson, P.and Gertzog, W.1982. Accomodation of a scientific conception: toward a theory of conceptual change, *Science Education* 66, pp 211-227.

Rozier, S. 1988. *Le raisonnement linéaire causal en thermodynamique classique élémentaire*. Paris, Thesis, Université Paris 7.

Rozier, S. and Viennot, L. 1991. Students' reasoning in elementary thermodynamics, *International Journal of Science Education*, 13 (2), pp 159-170

Schiele, B. 1984. Note pour une analyse de la notion de coupure épistémologique. In.Schiele, B., Belisle, C. and Garnier C. (eds) *Communication-Information: les représentations*. 6 (2-3). CIRADE, Montréal pp 43-98.

Shipstone, D. 1985. Electricity in simple circuits. *Children's Ideas in Science*. Open University Press, Milton Keynes, pp 33-51.

Tiberghien, A. 1984b. Revue critique sur les recherches visant à élucider le sens des notions de circuits électriques pour des élèves de 8 à 16 ans, in *Recherche en Didactique de la Physique, Les Actes du premier Atelier International: la Londe les Maures 1983*, CNRS, Paris, pp 91-107.

Tiberghien, A. 1984c. Revue critique sur les recherches visant à élucider le sens des notions de température et chaleur pour des élèves de 10 à 20 ans, in *Recherche en Didactique de la Physique, Les Actes du premier Atelier International: la Londe les Maures 1983*, CNRS, Paris, pp 55-74.

Valentin, L. 1983. *L'univers mécanique*, Hermann, Paris.

Viennot, L. 1989b. Obstacle épistémologique et raisonnement en physique: tendance au contournement des conflits chez les enseignants. in N.Berdnaz et C.Garnier (eds) *Construction des savoirs, obstacles et conflits*. CIRADE, Montréal, pp 117-129.

Viennot, L. 1993. Temps et causalité dans les raisonnements des étudiants, *Didaskalia* 1, pp 13-27.

Part two

The impact of common sense
Some investigations

The studies presented here give further details on the main lines of common sense reasoning identified in the first part of this book. Each of these studies can be read independently. The reader who is pressed for time can, at any point, move on to the conclusion without missing the gist of the message.

Chapter 6

Quantities, laws and sign conventions

1. INTRODUCTION

Common thought deviates from physical theory, and may lead to erroneous predictions about phenomena, due to its overly realistic approach to situations studied in physics, and to the intrinsic characteristics it attributes to physical quantities.

These aspects also affect how students deal with sign conventions and algebraic quantities.

2. THE ESSENTIAL: ALGEBRAIC QUANTITIES AND LAWS

As has been pointed out earlier, physical quantities are constructs of physical theory, and it is important to specify the way in which they relate to the phenomena under study. The sign convention adopted is therefore essential. For what we call algebraic quantities, choosing a sign convention means allocating the signs + and –: the numerical values of such quantities are what mathematicians call real numbers, which may be positive or negative, for example +2.7 or –2. Of course, the unit is indispensable, e.g., +2.7 ampere, - 2 metre, etc. But that's another story. Here we shall focus on the signs, which, it should be noted from the outset, also emerge as an issue when writing laws in the form of algebraic equations.

What does one need to know about + and – signs in order to practise physics?

2.1 Algebraic quantities

In order to specify the sign which makes it possible to assign a real number to an observation, it is very often necessary to have an axis of reference. Depending on the orientation of this axis, the same physical situation is represented by quantities with opposite numerical values. Thus, a position on a rectilinear path can correspond to x=+300m or x=-300m, depending on the orientation chosen on an axis parallel to the path, and on the chosen origin. Or an electric current can be characterised by the statement i=+25mA, or i=-25mA, depending on the direction chosen for the currents that one decides are positive. Strictly speaking, the name of the symbol has to be changed if the sign convention is changed – a point we shall return to later.

Once a physical quantity has been defined (i, for example), its numerical value depends on the physical situation – in this case, the effective direction in which the charges are moving in the wire. If the electrons change direction, the current changes “sign”.

To specify the “sign” of a quantity is to say whether its numerical algebraic value is greater or less than zero. In other words, the “signs” of quantities speak of an inequality: when one writes “i=+25mA”, the “+” means “i>0”.

In short, whether the value of a quantity is positive depends on two things:

- the chosen sign convention
- the particular physical situation.

2.2 Algebraic relationships between quantities

The question this time is to determine what sign to place between the symbols representing the quantities. In the case of an electrical component, for example, should the relationship between potential difference, resistance and current be written as $V_A-V_B=+Ri$ or as $V_A-V_B=-Ri$? The previous question, which concerned the sign *of* (...a numerical algebraic value) becomes the question of the sign *between* (...algebraic quantities).

Since the definition and the algebraic value of quantities depend on the chosen sign convention, the relationship cannot but depend on it too. But once the sign convention has been decided on, and the definitions fixed, there can be no “special cases” – the law must hold for all situations, without exception. If this were not so, how could physics be practised? Laws and equations cannot change with each new situation. Therefore, the answer to the question “Should there be a + or a – in this relationship?” depends only on the sign convention.

These essential points are outlined in box 1.

Box 1
The foundations of algebraic formalism: + and – in physics

Two questions

1. Is the algebraic value of a quantity positive or negative?
(Sign "in" the algebraic quantity; for example, i=+4mA or i=-4mA?)
2. Should a relationship be written with a + or a – between physical quantities?
(sign "between" the quantities in the relationship; for example $V_A-V_B=+Ri$ or $V_A-V_B=-Ri$?)

Does the answer depend on...

...the sign convention chosen?

e.g. :

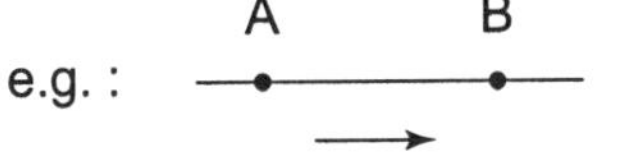

direction of the currents one has decided are positive

...the particular physical situation?

e.g. :

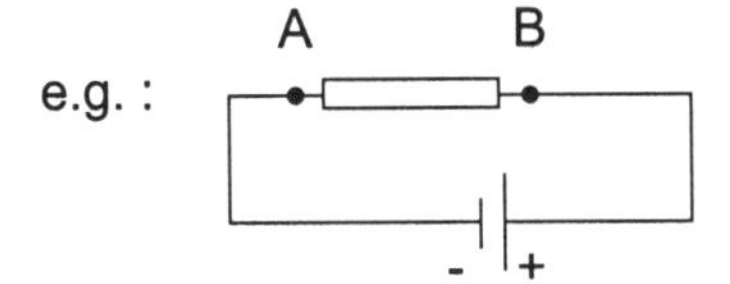

...the sign convention chosen?	...the particular physical situation?
1. the answer depends on the sign convention	1. the answer depends on the situation
2. the answer depends on the sign convention	2. the answer does not depend on the situation

3. THE NATURAL: REALITY FIRST, LAWS MUST ADAPT

It is natural to think of the world in realistic, and therefore "positive", terms. Our vocabulary often has two words to express the opposite poles of a single notion – ebb and flow, height and depth, etc. – whereas algebraic language uses only one symbol (e.g., Φ: flux, z: altitude). The dual designation of notions makes it possible to associate all the words of ordinary language with what are known in algebraic formalism as positive values (it is more natural to say: "at a depth of 10m" than "at an altitude of –10m"). It is said that nature abhors a vacuum; we might also say that common thought is none too fond of a "minus."

The problem is that when one insists on having "pluses", one continually has to fiddle with relationships. For example, the relationship between the current i "arriving" on the plate of a capacitor and the charge q on the plate in question can be written i=dq/dt (dq/dt is the rate of change of charge with time). But if one's heart is set on keeping the current i always positive, the discharge must be written as i=-dq/dt, because when the charge decreases, the quantity "dq", associated with the variation of the charge, is negative; and for it all to "work out," there has to be a minus sign between the symbols. And so we have a relationship that depends on the situation: if the circuit is oscillating, the alternating current makes it necessary to change the relationship for each half cycle!

Moreover, there is an evident, albeit less striking, tendency to associate two notations, for example "+q" and "q>0". The "sign *between*" (in this case, before) and the "sign *of*," in other words the sign expressing a relationship and the sign expressing a value, are combined here. If this "q" has got to be positive, then the two notations are equivalent; but then there is one too many: why place a "+" sign before q? For no reason, except that, if the charge is negative, it has got to be shown somehow. Enter the "minus" sign, and according to this logic, "-q" means "caution, negative charge". People prefer to put a "minus" sign *in front of* a physical quantity rather than imagine it *inside* its value. But then more problems arise. If, for example, one wants to express the potential difference V_A-V_B between the ends of a capacitor of capacity C, one plate of which has charge q, it is necessary to write $V_A-V_B=q/C$ or $V_A-V_B=-q/C$, depending on whether the charge is positive or negative. Unless one opts for the cover of ambiguity, and states, as students are wont to: "The equation becomes negative". Does that statement mean that a "minus" sign should be inserted *between* the quantities involved in the relationship, or that the term on the left has a value that is lower than zero? Asking oneself this question – merely understanding it – runs counter to natural tendencies of thought.

This makes for a great many difficulties in the use of algebraic formalism. Here "formalism," to a much greater extent than in the preceding chapters, is to be taken in the usual sense of the term: symbols, equations, or "formulae." Can the natural possibly intrude even in this most artificial and academic of domains?

4. RESULTS OF INQUIRIES[1]

It certainly looks like it. The statements above are grounded largely on the results of inquiries involving pupils in the last years of secondary education or university students, concerning what are generally held to be extremely simple physical situations. Box 2 concerns one of the "most

[1] For more details, see Viennot (1980, 1987); also, Rebmann and Viennot (1994).

popular" relationships of elementary mechanics: the phenomenological relationship expressing the behaviour of a linear spring: F=-kx (where F is the algebraic value of the force exerted by the spring on an object attached to its end, x is the extension of the spring, along the same axis, and k is the force constant of the spring.

Box 2

Understanding an algebraic relationship – Rebmann and Viennot (1994)

Summary of the question

Consider a spring with negligible mass[a]; at each end of the spring, an object is attached. The spring exerts on the objects (at each end) the forces **F**1 and **F**2 respectively, with magnitudes $\|\mathbf{F1}\| = \|\mathbf{F2}\| = k|\Delta l|$,

Δl is the extension of the spring (length – unstretched length)

k is the spring constant of the spring.

The force **F** exerted by the spring on an object attached to one end of it is drawn on each of the following diagrams.

In each case, O is the origin of the x axis.

For the second case, write the algebraic value of **F** on **Ox** (F) using the same symbols as those used in the diagram and the constant k.

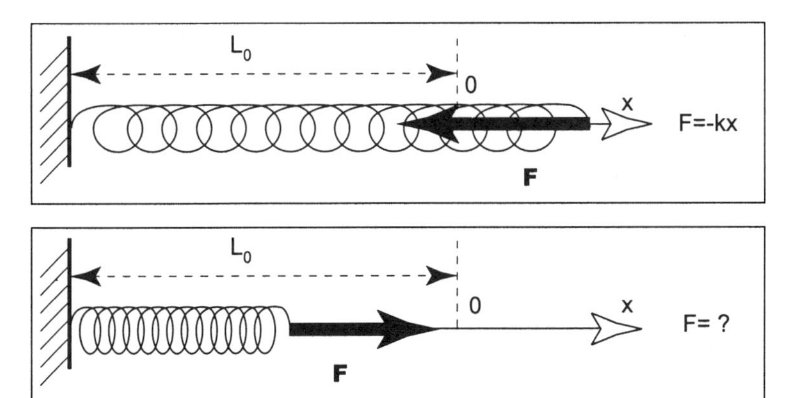

A common error:

Writing, for the second case, F=+kx (probably because F is then positive).

Population	Number	University year	Answer "F=+kx"
Physics students	N = 306 N = 87	1st 3rd	45% 40%
Teachers in a training session	N = 49	4th	33%

a. More precisely, one should specify "relative to the mass of the other objects under consideration".

The results reveal that an astounding proportion of students write the relationship as F=+kx when the spring is compressed, probably in an attempt to express the positive direction of the force. Faced with a problem about oscillations, the same students would have had no difficulty using the correct relationship. But in the case of the question in box 2, they are asked to think about what they are doing. All it takes is a graphic representation of the situation, an arrow drawn in a given direction, for realism to take over and for "positive" reasoning to overflow, sweeping aside the permanence of algebraic relationships.

Box 3
Two versions of Kirchhoff's first law – see Viennot (1983, 1987)

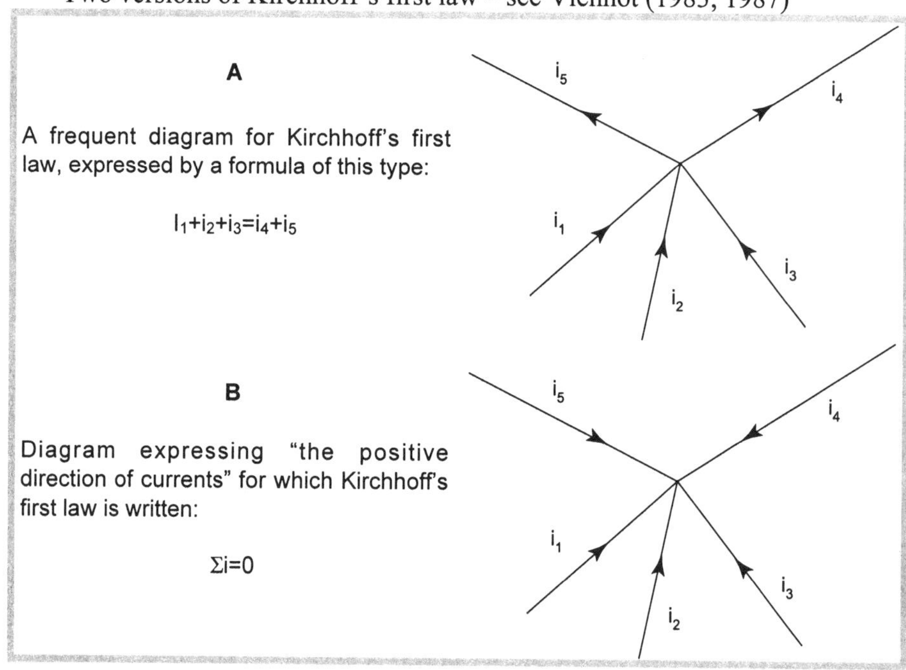

5. REALISTIC BALANCES

The need for a good understanding of algebraic formalism is particularly clear as regards the writing of balances in physics. These are relationships expressing laws of conservation. For example, Kirchhoff's first law, in electricity, expresses the conservation of charge in a steady state situation (or in quasistationary change: see chapter 5). This law is sometimes written as in box 3 (A). All the quantities represented are implicitly positive. In other words, the "positive" directions are implicitly identified as the actual directions of the currents in each wire. Once the law has been written in this form, it is not too difficult to accept and to interpret negative values of current. So this form of expression is both natural and legitimate: it stems from a realistic vision and adapts itself to various situations.

Box 4

Two ways of writing the conservation of energy for one cycle of a heat engine – Viennot (1983, 1987)

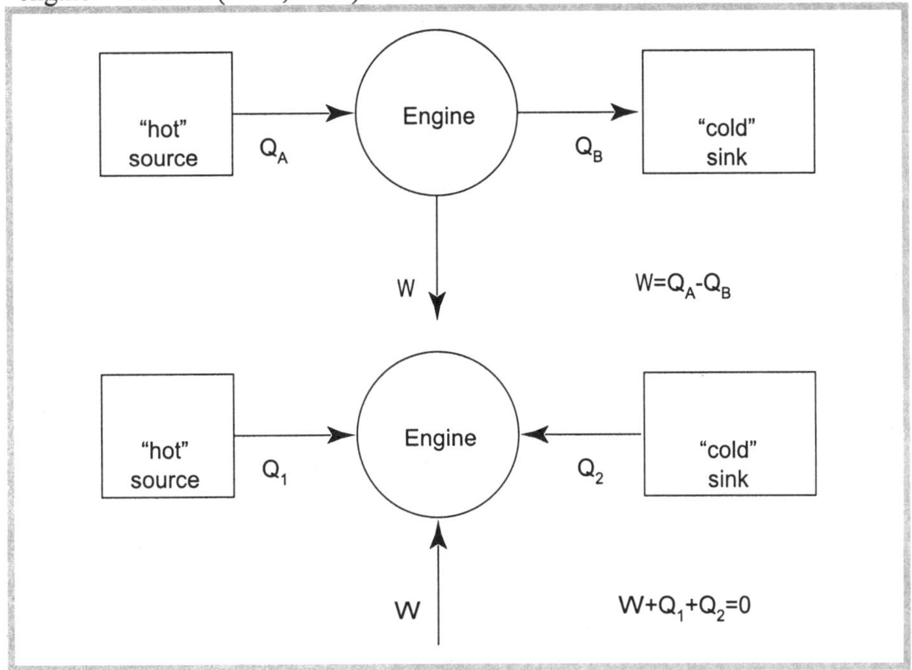

But this way of expressing charge conservation is not the one that applies most generally, if only because it considers a particular number of branches for a given node. By contrast, the relationship $\Sigma i=0$ is not only general but also symmetrical. The "positive" directions are all towards a node, or all away from a node. Therefore, the corresponding diagram for, say, 5 branches, would look like the one in box 3 (B). Such a representation is not often found in textbooks, even at university level, probably because a diagram of this type cannot represent a realistic situation. And its true status then has to be admitted: it is a defining diagram, not an illustrative one. On the other hand, the diagram in figure 1 allows both kinds of status; so no choice between them ever has to be made.

These remarks can be applied to other types of balances, for example, to those involving the conservation of energy for heat engines. Box 4 presents all the elements implied in the law for behaviour at a node. The difficulty of accepting a diagram with converging arrows, for example, can be interpreted in the same way as above: no engine can complete a cycle if it only ever receives energy. Just as in the case of the diagram where all the arrows converge towards a current node, it is impossible to ascribe a realistic meaning to this diagram. That is the price we pay for a general and symmetrical way of writing laws.

6. A SUGGESTED STRATEGY: SPLIT DIAGRAMS

There is no reason why we shouldn't try to lower the "cost" of a conceptual tool. If we want both to facilitate the use of algebraic formalism and to give access to the mechanisms underlying it, it is probably essential to tackle the problem head on.

Box 5

Suggestion: sign convention and realistic representations for a heat engine

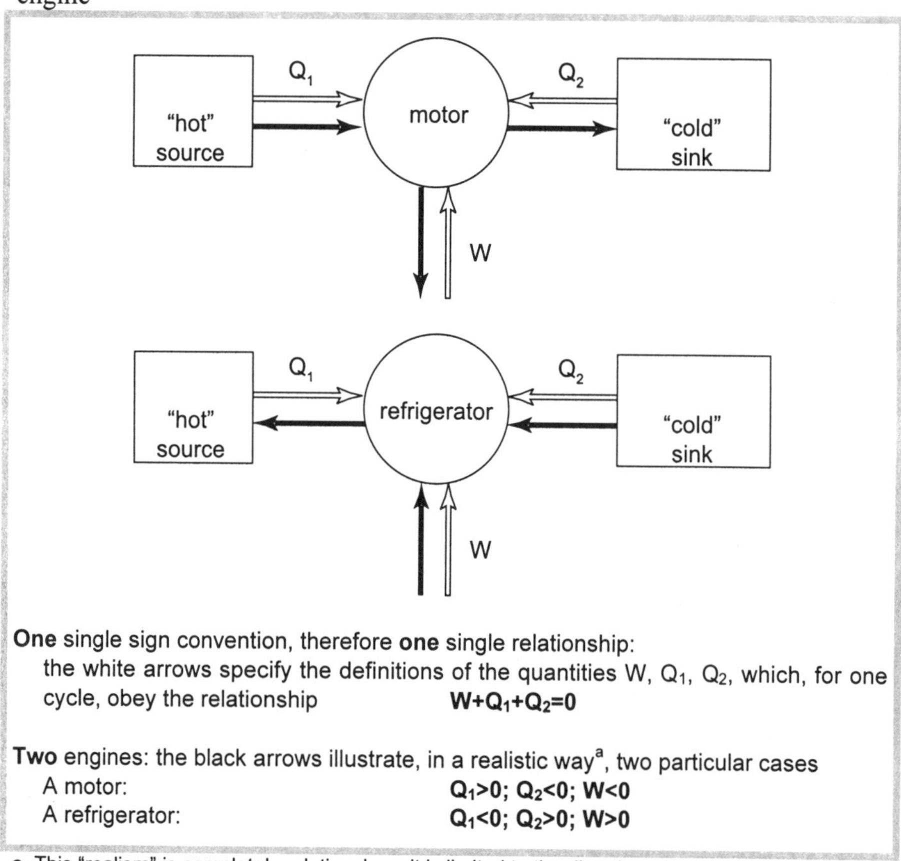

One single sign convention, therefore **one** single relationship:
the white arrows specify the definitions of the quantities W, Q_1, Q_2, which, for one cycle, obey the relationship $\mathbf{W+Q_1+Q_2=0}$

Two engines: the black arrows illustrate, in a realistic way[a], two particular cases

A motor: $\mathbf{Q_1>0;\ Q_2<0;\ W<0}$

A refrigerator: $\mathbf{Q_1<0;\ Q_2>0;\ W>0}$

a. This "realism" is completely relative; here it is limited to the direction of the flows of energy.

We suggest that the two types of elements in the diagrams be explicitly separated. In one part of the drawing, the arrows defining quantities, as in the diagrams with converging arrows in the examples given above. They are drawn in white in box 5. The arrows describing a particular situation in a semi-figurative way should be drawn separately, in a different manner; they are drawn in black in box 5. Each arrangement of white arrows corresponds to a single sign convention, and therefore to a single form of the algebraic expression of the law.

The various situations may then differ in the numerical value of the quantities; only the signs are discussed here. The sign can be easily seen in the split diagrams presented above: black and white arrows pointing in the same direction indicate that the value is positive. If the arrows do not point in the same direction, then the value is negative.

Such a technique may seem childish – after all, this isn't kindergarten! We'll have to bring in our coloured pencils next (as we've used up black and white)!

But why should this offend anyone? Our aim is to show that the formal tools we have used are not all in the same category; nor do they have the same meaning. We recommend using colours in diagrams showing the interactions between objects (chapter 10); and in diagrams representing balances, they clarify things so well that they are almost indispensable. Verbal statements are not sufficient in these cases, any more than in the study of forces.

7. VERBAL STATEMENTS: CONFUSION AND MISUNDERSTANDINGS

The three sentences below were taken from the same school text, dealing with the oscillations of a mass-spring system that is damped by friction, γ being a characteristic coefficient of that friction, and v the velocity of the oscillating mass. About the oscillator, in the space of three pages the reader is told that:

- the power transferred[2] by friction is $-\gamma v^2$;
- the power dissipated through friction is γv^2;
- the power dissipated through friction is $-\gamma v^2$.

There is no consistency in the designations and algebraic expressions applied here. Indeed, the third statement contradicts either the first or the second one, depending on whether one decides that the verbal expressions are correct but not the algebraic ones, or vice versa.

[2] In French: "reçue"

What happened when the writer was putting all this down?

Status became confused. Let us associate colours with the two registers in which one can make verbal statements. Box 6 shows how these sentences can be made compatible with physical theory.

The first designation is a definition: that of an algebraic quantity – "the power transferred from..." (shown in white), which is positive when the energy of the system increases. This is not the case in the particular situation being considered, where the mass-spring system loses energy. So the value of the quantity is negative ($-\gamma v^2$ is always negative).

Box 6

Words and energy flows – Viennot (1983, 1997)

Energy dissipation in three (interchangeable?) statements – for a mechanical oscillator with damping coefficient γ and with a single mobile mass considered, at speed v.

1. the power transmitted by the environment to the oscillator is $-\gamma v^2$
2. the power dissipated through friction is γv^2
3. the power dissipated through friction is $-\gamma v^2$

- Statement 3 contradicts statements 1 and 2.

- In order to give each statement a precise meaning, closer, presumably, to the one intended by the author, one can decide that
  - a verbal expression underlined in "white" designates an algebraic quantity defined by a white arrow (on the diagram)
  - a verbal expression underlined in "black" represents the effective direction of the energy transfer (also on the diagram).

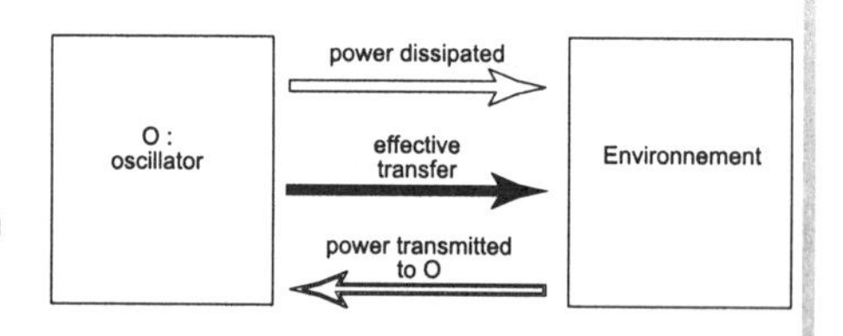

- In each of the statements 1 and 2, the verbal and algebraic expressions of the quantity used are consistent.
- Statement 3 remains iffy: a realistic verbal expression ("dissipated", underlined in black) is associated to $-\gamma v^{22}$ the algebraic expression of a quantity defined in "white": "received power."

The second one is also a definition (in white): "dissipated power" is a quantity that is positive when the system loses energy. But the expression can also describe the phenomenon in a realistic fashion (in black): some power is indeed dissipated, to use ordinary language. No problem there, everything is positive.

In the third statement, the two registers are fused: the verbal expression is realistic (in black) but the algebraic expression is the value of the received power, as defined (in white) in the first statement.

8. CONCLUSION

We may not be shocked by the more or less random use of these algebraic terms and expressions when we are already conversant enough with the subject being discussed to make all the necessary adjustments as we go along, including switching from one language another in mid-sentence. In other words, when we do not need the information. That is why most professional physicists would not notice any discrepancy in the three preceding statements. They already know.

But what about our students?

When balances are established, it is apparent that the difficulties mentioned here are an obstacle to understanding, and the method suggested above seems all the more necessary.

Is it good teaching practice to continue using statements that half of the students do not understand? Let us take the example of the well known "physicist's convention", summed up as "the work done *on* the system is counted positive": it is, in fact, a definition, comparable to the one for received power (in white). First of all, we would be well advised to rename this "convention"; it is a definition. Then, it is often incorrectly understood to mean "the work done *on* the system is always positive." If that statement were true, the work done *on* the system would no longer be an algebraic quantity but an element of a realistic description (in black).

In short, "positive counting" can be very ambiguous: is the sign that *of* the numerical quantity ("The work done *on* the system is always positive"), or that *before* the quantity in a literal expression ("The work done *on* a system, W, and the heat transferred *to* it, Q, are such that the change in energy of the system, ΔE, is given by the expression $\Delta E=+W+Q$, regardless of the signs of these quantities)? Ambiguous formulations of this sort are probably due to a refusal to distinguish between the terms of such an alternative, or even to understand the implications of the question that is raised.

Words are too misleading, too loaded with reality, for us not to seek help elsewhere. And so – enough pretentiousness, out with the coloured pencils!

And yet, it must not be forgotten that the surest guide to pedagogical action is good sense. No recommendation should be taken as dogma. The clarification procedures suggested here must not be injected into teaching in a continuous, absolute, and ultimately stultifying fashion. However, one can envisage a progressive approach, with stress laid from time to time on pedagogical activities that make algebraic formalism meaningful.[3] And in any case, it is hard to disagree with the view that, at some point in their studies, students of physics should understand what they are doing when they manipulate the most elementary algebraic expressions.

REFERENCES

Rebmann, G. and Viennot, L. 1994. Teaching algebraic coding. *American Journal of Physics*, 723-727.

Saltiel, E. and Viennot, L. 1983. *Questionnaires pour comprendre*, Université Paris 7 (diffusion LDPES).

Viennot, L. 1983. Pratique de l'algèbre élémentaire chez les étudiants en physique. *Bulletin de l'Union des Physiciens* 622, pp 783-820.

Viennot, L. 1987. Recherche en didactique autour de la transition Secondaire-Supérieur. *Bulletin de l'Union des Physiciens* 699, pp 1251-1268

[3] The accompanying document for the grade 11 syllabus (Première scientifique) produced by the Physics GTD (March 1994) proposed this approach to sign conventions at the national level for the first time. See also Saltiel and Viennot, 1983.

Chapter 7

Changing frames of reference at eleven

From Maury, Saltiel and Viennot (1977)

The help received from Edith Saltiel in the preparation of this chapter is gratefully acknowledged.

1. INTRODUCTION

In situations involving changes of frames of reference, students adopt distinctive forms of reasoning which can account for many of their difficulties (chapter 3). Do such forms of reasoning appear at an early age, and, in particular, before any instruction on that particular point?

The average age of the children participating in the study summarised below is eleven; they are in the last grade of primary school.

2. THE EXPERIMENTS: PRINCIPLE AND DESCRIPTION

The following aspects of reasoning about motion are common among adults:

One is a difficulty in dissociating the kinematic and geometrical aspects of motion. The initial and final positions of the moving object are often considered as material points that are fixed in space, and the time factor is eliminated.

Another difficulty is linked to the cause of motion: the speed of an object and the direction of its displacement are primarily seen as resulting from the action of a motor, rather than from measurements in a given frame of reference. This explains why, in common arguments, the speed and the direction of displacement relative to the motor appear as quantities that intrinsically characterise motion.

Three parallel experiments were devised for children. They share a common framework, but each one focuses on a separate "sensitive aspect" of motion.

In an individual interview, the child is asked to give his/her opinion on two observations concerning a single event, each from a different frame of reference: are they compatible, and, if so, under what conditions? The two frames of reference in question are the room and the child, who, sitting "stock still" in a chair on castors, can move through the room laterally. The episode that is analysed is the lateral displacement of a pencil[1] in front of the child, who observes it through a tube that limits his/her field of vision.

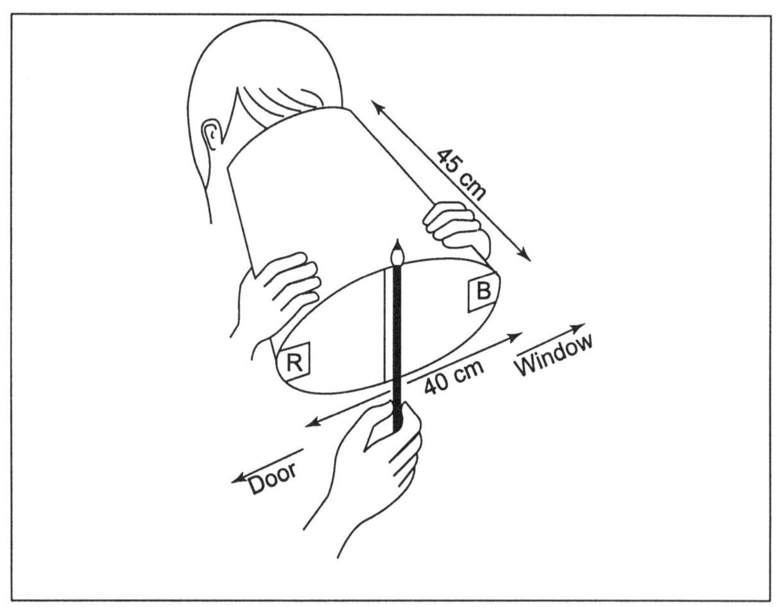

Device used to observe the pencil

[1] This experimental device has one advantage: neither the pencil nor the chair imply through their geometry a privileged direction for the lateral displacement. When analysing the motion of a vehicle, on the other hand, the fact that it has a "front" and a "rear" is likely to reinforce the idea of an intrinsic direction of displacement. This probably explains why, for fourth-year university students, there is still a surprisingly large proportion of answers along these lines in a problem concerning a boat overtaking another boat (75%, Saltiel and Malgrange, 1979).

The direction of the motions observed are described, within the room, as towards the door (to the right of the child) or towards the window (to the left); in the tube, there is a red mark (on the right) and a blue one (on the left).

The idea is to stage situations in which the child always sees the same thing: the pencil moving towards the blue mark. This observation is compatible with several displacement combinations for the pencil and the chair within the room. Table 1 shows the cases in which this type of observation is possible.

Each case corresponds to one phase of the interview. The first, where the chair/child unit is in a fixed place in the room, consists in presenting to the subject the observation which is the focus of the experiment, and explaining the motion markers described above. In the three other cases (situations I, II, III), the child is questioned as follows:

Table 1
Various combinations of motion in which the observation made by the child remains the same

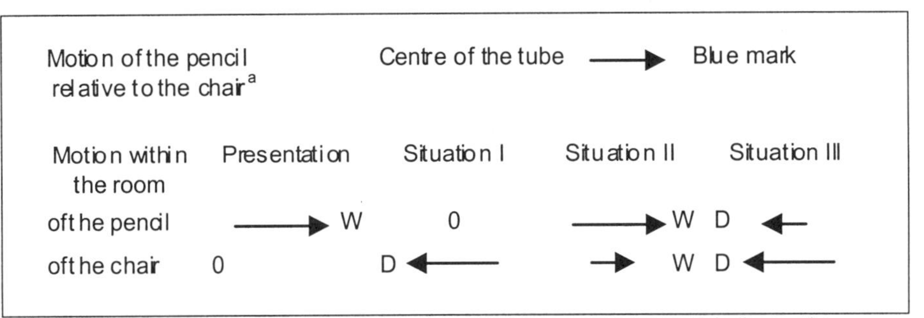

Motion of the pencil relative to the chair[a]	Centre of the tube ⟶ Blue mark			
Motion within the room	**Presentation**	**Situation I**	**Situation II**	**Situation III**
of the pencil	⟶ W	0	⟶ W	D ←
of the chair	0	D ⟵	→ W	D ⟵

a. The arrows indicate the direction of the displacement; their length indicates the relative values of the speeds. 0 indicates that the displacement is zero. D: towards the door. W: towards the window.

"Would you be able to see this (displacement of the pencil from the centre of the tube towards the blue mark) if…

(Situation I) …the pencil were not moving within the room?

(Situation II) …the pencil and the chair were moving towards the window in the room?

(Situation III) …the pencil and the chair were moving towards the door in the room?"

If the child answers *yes* for situations I and II, he/she is also asked:

"Is this possible no matter what the speeds of the pencil and the chair are in the room? If not, does one of them have to move faster than, more slowly than, or as fast as the other?"

Thus, a prediction is asked for in each case. If it is wrong, or if the child hesitates or gives no answer, the situation referred to is acted out – not to convince the child, but to observe his/her reactions.

Three versions (A, B, C) of this experiment were carried out.

- In order to evaluate the effect of direct perception of the direction of motion, there is no visible motion in experiment B. That is the only difference between experiments A and B. In A, the child observes the continuous displacement of the pencil from the centre of the tube towards the blue mark; whereas in B, a mask is used, and the child observes only the initial and final positions of the pencil, for the same displacement.
- Experiment C aims at determining the importance in common reasoning of the motor which causes motion. In fact, experiments A and C differ only in that, in A, the experimenter moves the pencil, whereas in C, the child does it.

72 children at the same grade level were divided at random into three groups, for experiments A, B and C respectively (A, parts I and II: N=15, part III: N=22; B: N=22; C: N=22).

3. MAIN RESULTS AND DISCUSSION

3.1 Overview of the results

There is a high proportion of correct answers for situation I in all three experiments (A: 87%, B: 73%, C: 86%), with some children giving a correct answer straight away, and others coming to an agreement with the experimenter after the demonstration. It is, therefore, readily accepted that if the observer moves in a given direction (relative to the room), a fixed object will appear to be moving in the opposite direction. This can be associated to the idea of "apparent motion" which adults often introduce in connection with situations in which a tree is seen from a moving train or a poster from a moving walkway (chapter 3). In such cases, where obviously "fixed" objects are commonly seen in motion, the usual interpretation seems to be that this is merely an illusion; no further attempt is made to attribute a dynamic cause to the phenomenon.

Difficulties arise when a condition has to be determined, in situation II: the pencil moves faster than the chair inside the room (both motions having the same direction as the apparent motion). In situation III, it also seems difficult for the child to admit that a motion perceived as having one direction is compatible with the motion of the pencil and the chair in the opposite direction, relative to the room (in that frame of reference, the object observed moves more slowly than the person observing it – see table 1).

As regards these difficulties, it would seem that the differences in the parameters of experiments A, B and C have specific effects.

3.2 Geometry and kinematics

Experiments A and B differ in that a motion which is perceived as continuous in experiment A is perceived only through initial and final positions in experiment B. These discrete data should, in theory, make it possible to reconstitute motion, particularly its direction. And yet the children react very differently to the two experiments. In experiment A, for situation III (where the motions of the observed object have different directions relative to the room and relative to the child), objections are numerous (32%) and persistent (22%): "The pencil cannot move both in this direction (→) and in that one (←)." Such objections are less frequent for experiment B, and they all break down when verbal stress is laid on the relative final positions of the moving object and of the observer inside the room: "Is it possible that, at the end, the pencil is farther away from the door than you are?" This procedure is far from being as effective for experiment A.

In each of these experiments, therefore, the children seem to focus on a different element, which they take to be an invariant. In experiment B, the relative positions of the moving object and the observer at a given point do, in fact, constitute a good Galilean invariant, on which the children can rely and achieve success. In experiment A, the direction of the motion, which is the main feature of the observation, is wrongly considered to be an intrinsic characteristic of the displacement of the pencil. This brings us to a second aspect of reasoning which needs to be considered.

3.3 The importance of a motor

As has been shown in chapter 3, the role of the motor is often seen as decisive in common analyses of motion. The fact that it is difficult to accept that a pencil can move "both in one direction and in another" (experiment A, situation III) may well be related to the idea that "it has been moved" in a given direction.

This becomes clear if we compare experiments A and C.

In experiment C, situation II provides a few clues. The motions all have the same direction relative to the room, and the child is asked: “What moves faster inside the room, you or the pencil?” Some of the answers are perplexing:

“That depends on the hand movement, on whether it is fast or slow;”

“The pencil moves faster because my hand moves faster;”

“It would be a coincidence if we both moved at the same speed.”

In fact, once the direction of the movement and the other motions have been decided on, there should be no “that depends”: the pencil always moves faster inside the room than the chair does. But it seems that for these children the expression “the speed of the pencil inside the room” is taken to mean “the speed of the motion,” i.e., “the speed supplied by the motor.” The experimenter’s remarks to the “note-taker” inquiring about the speed of the pencil have no effect – which is not surprising, if we recognise that children do not ask themselves the question: “Speed with respect to what?” Focusing on the motor, they evaluate the only speed that they take into consideration, the speed of “the hand movement”, which they attribute to the pencil once and for all. In the language commonly used by adults who have had some instruction in physics, they consider the pencil’s “intrinsic” speed.

It is also understandable that, in experiment C, situation III should cause even greater turmoil than experiment A. This time, 86% of the children begin by rejecting the possibility that the pencil moves “both in this direction and that one.” The direction of the motion of the pencil is no longer merely observed, it is “caused” by the child, and is thus readily ascribed the status of an intrinsic characteristic. Surprisingly, such resistance is short-lived, and all of the children are, in the end, convinced that what first seemed to be contradictory motions for the pencil are in fact compatible. This is probably due to the fact that situation C is a “drag” situation, in which two causes, namely the displacement of the chair and the motion of the hand, are added up algebraically. This is not the case in experiment A, where a third of the children come to a standstill in situation III: no dynamic cause can be directly associated to the perceived reverse motion, whereas the moving object is propelled in the opposite direction by a clearly identified motor.

4. CONCLUSION

This study confirms our initial hypothesis: the traits of common reasoning found in university students on the subject of frames of reference are already present in the reasoning of children. The geometrical and kinematic approaches are not reconciled, and there are confusions between a purely descriptive (kinematic) point of view and a causal (dynamic) one.

If we want to make a proper mastery of basic kinematics possible for our pupils, it seems important that we should lead them to disconnect motion from the idea of a motor. It is, of course, necessary to highlight the key expression, "relative to," but it is also necessary to establish a relationship between initial and final positions on the one hand, and the direction of a motion on the other. Exchanges of the kind used in these experiments, when adapted to our pedagogical aims, can be useful teaching tools. More generally, this can be seen as an opportunity to oppose a formalism which, though simple, can apply to many situations, to the sometimes contradictory intuitive aspects that stand in its way. Mastering the different situations in table 1 is in itself a major conceptual breakthrough. The point is to introduce the idea that accepted physical theory leads to a unified understanding of events, and makes non-trivial predictions possible.

However, caution is necessary: the study of changing frames of reference is fraught with difficulty, and should be broached only after a careful adjustment of ambitions and constraints. Placing the pupil in a given "physical" situation may prove to be a decisive factor – in fact, it may well prove indispensable.

REFERENCES

Maury, L., Saltiel, E. and Viennot, L. 1977. Etude de la notion de mouvement chez l'enfant à partir des changements de repère, *Revue Française de Pédagogie*, 40, pp 15-29.

Saltiel, E. and Malgrange, J.L. 1979. Les raisonnements naturels en cinématique élémentaire. *Bulletin de l'Union des Physiciens*, 616, pp 1325-1355.

See also the references for chapter 3.

Chapter 8

Common reasoning about sound

In association with Laurence Maurines. This chapter makes use of the results of a study by that author, and is largely based on her article in *Trema* (Maurines, 1993).

1. INTRODUCTION

The investigation outlined here studies common reasoning in acoustics. It is the continuation of research analysing the difficulties encountered by pupils in the study of the propagation of a transverse signal on a rope.[1] Are the mechanistic types of reasoning that are at the root of some of these difficulties also found in connection with non visible signals, propagating at appreciably higher speeds? Some elements which may provide an answer to this question are given below, following a few notes on how "travelling bumps" on ropes are commonly analysed.

2. PROPAGATION OF SIGNALS IN SECONDARY TEACHING

The subject of this research was suggested by the syllabuses for grade 11[2] dating from before 1994, when the study of waves was first introduced.[3] The

[1] See Maurines (1986), Maurines and Saltiel (1988a); see also chapter 4.
[2] Première S, sixth year of secondary education in France (science section).
[3] In 1993, sound was included in the syllabus for grade ten (i.e., Seconde, the last year of undifferentiated secondary education in France – two years before the baccalaureate).

approach was mainly experimental, and centred on macroscopic modelling. The pupils were presented with a series of experiments on the propagation of various types of signal (on a rope, a spring, or water, sound signals and luminous signals), to be studied along with graphs corresponding to two descriptions (see box 1):

- the spatial description, representing the state of the propagation medium at each point in space, at a given instant;
- the temporal description, representing the changes with time of the state of the medium, at a given point in space.

Box 1
Spatial and temporal descriptions for a bump propagating on a rope.

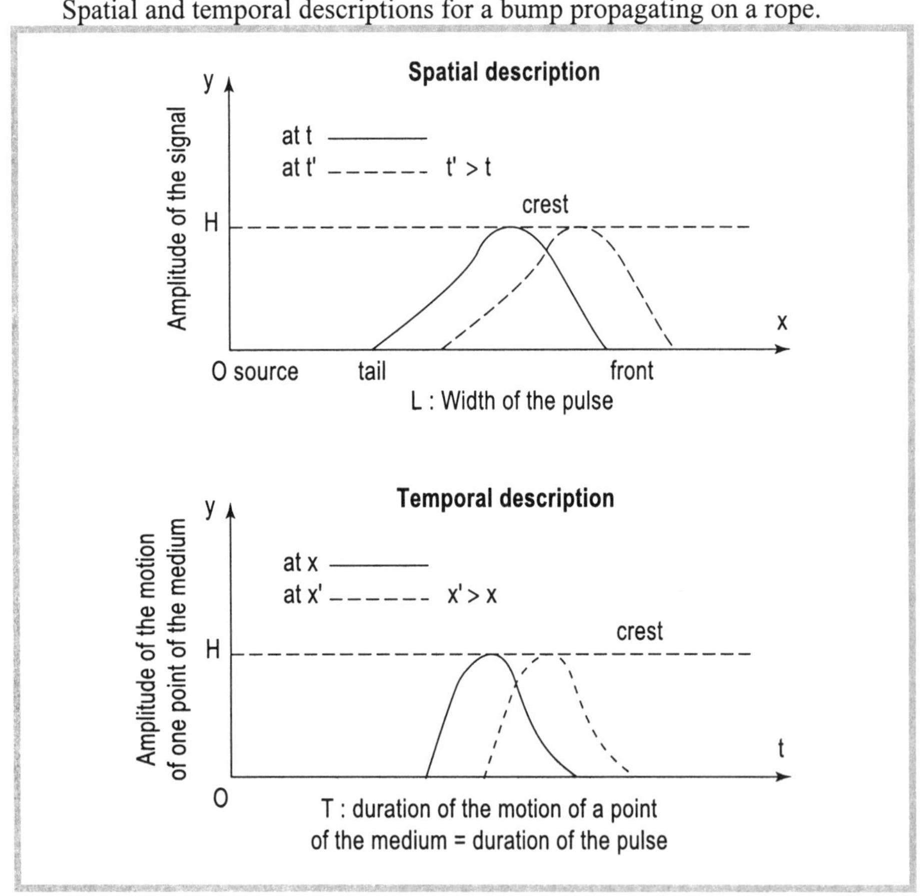

This graph-based approach was devised primarily for the propagation of a pulse along a rope, in cases where the medium is considered "perfect" (a one dimensional signal propagates without being deformed).

The goal was to teach the pupils

- to identify the source, the medium, and the possible presence of friction or other phenomena that dissipate energy;
- to associate with the phenomenon the quantities "speed of propagation" (also called "wavespeed"), "duration of the signal", "width of the signal," and "amplitude of the signal";
- to analyse what the quantities depend on and do not depend on; in particular, they must recognise that the wavespeed depends solely on the medium and on its physical characteristics, and not on the amplitude or shape of the pulse (when the medium is "linear" and "non dispersive," though that restriction is not made explicit at this level);
- to relate correctly the spatiotemporal description of graphs to a phenomenon involving propagation.

At the university level, there is a further teaching objective: mastery of the mathematical formalism of the equation of waves and of its solutions.

3. MAIN RESEARCH FINDINGS ABOUT PULSES ON ROPES

At each of the grade levels studied – from grade ten, prior to instruction, to the third university year, after instruction[4] – the characteristics of reasoning described earlier in this book (chapter 3) recur with remarkable frequency.

A considerable proportion of the pupils questioned before instruction on the subject (60%, N=42) and afterwards (75%, N=16) say that, for a given rope, the speed of the propagation of the bump depends on the hand movement which caused it. Many dynamic justifications are provided with the answers:

"The bump will move faster and faster if the hand moves fast."

"The speed depends on the force with which he moved his hand."

Some comments even refer to "the force that is propagated," and on many diagrams a "force" is drawn on the travelling bump.

[4] Population questioned on this theme: approximately 700 pupils prior to instruction on waves (Seconde, Première scientifique, Première technique, Terminale technique) and 600 pupils and students following instruction on the subject (Première scientifique, Terminale scientifique, science students in the three first years at university).

The source therefore appears to be sending some "dynamic capital" to the rope when the signal is formed; this hybrid notion combines force, speed and energy.

Moreover, the nature of this "capital" is indicated by the shape of the bump. At least, this is what is suggested by the results of a questionnaire on the speeds of propagation of three signals of differing shapes on one rope. In fact, 87% of the pupils before instruction (N=93) and 41% after instruction (N=27) state that these speeds differ. Many of their justifications associate the speed and the amplitude of the signal:

> "[Bump] C moves faster, because the force supplied by the child's arm modifies the shape and the speed of the bump. Therefore, the more intense the force is, the bigger the pulse on the rope and the speed of the bump."

> "That depends on the force with which the movement was made. You can see it by the size of the bump. It reflects the force exerted by the child to arrive at that result."

The motion and the shape of the bump are, therefore, seen as two facets of the dynamic capital of moving signal. If friction affects this capital, both these aspects evolve together, according to the students. Indeed, a situation in which a bump disappears before reaching the other end of the rope gives rise to comments like:

> "The height diminishes, because the movement of the hand slows down."

> "If the bump disappears, that is because the force that made it has disappeared; at the same time, the speed diminishes".

68% of the pupils before instruction (N=56) and 55% afterwards (N=42) believe that there is a decrease in speed.

In short, the forms of reasoning observed often seem based on the notion that the signal receives some dynamic capital, a sort of object that the source provides, and that is materialised in the bump, and may run out. Consequently, the speed of the signal is not understood as depending solely on the characteristics of the propagating medium. There is a similarity between reasoning of this type and the forms of common reasoning identified in the mechanics of solids, where a dynamic supply due to the source, and located within the moving object, determines its speed, running out when there is an opposite force. The pupils identify the mechanics of a signal with the mechanics of a moving object.

One might expect to find the same confusion in connection with the propagation of sound.

4. PROPAGATION OF A SOUND SIGNAL

The speed of propagation of a sound signal depends only on the propagation medium; it possesses the property mentioned above for a bump on a rope: in a given ("linear") medium, it does not depend on the amplitude of the signal. Are inappropriate associations made between the speed of propagation and other factors in this case, too?

The study findings given below concern pupils questioned before any instruction on waves, in grades 9 and 10[5] (approximately 550 in all).

4.1 Speed of sound and power of the source

One question (box 2) involves a comparison of the speeds of two sound signals emitted by two sources with different power.

Box 2

Does sound propagate more quickly if we shout louder?

Summary of the question

P• M• J•

Three children, Peter, Mary and John, are on a straight road. Mary is at an equal distance from both Peter and John. At the same instant, Peter and John sing the note E. Peter sings louder than John. Does Mary begin to hear Peter and John at the same instant?

❑ YES Why?

❑ NO Why?

Correct answer

Yes; the speed of a sound depends solely on the medium ("linear" situation), and the distances covered are equal.

Results

At grade 9 (N=62), before any instruction on propagation:

- 40% of the pupils answer as if the two sounds propagated at different speeds: "She hears Peter first." 39% attribute this to the power of the source: "...because he's singing louder."
- 42% give the correct answer, YES.
- 18% give no answer.

[5] Troisième and Seconde (France).

For 40% of the pupils questioned (N=62), one does not begin to hear two sounds emitted simultaneously by two sources at the same instant, even though the sources are equidistant. According to these pupils, the source with the greatest power is heard first. Their comments establish an explicit link between the amplitude of the sound and the speed of its propagation:

"She hears Peter first, since he's singing louder."

"She hears Peter first: since the sound is louder it is projected more quickly."

4.2 Speed of sound and amplitude of the signal

The tendency to link the amplitude and the speed of propagation of the signal reappears in the results obtained for the questions outlined in box 3. The pupils are asked to compare two sounds emitted by identical sources; one sound propagates in the open air and the other one is channelled by a hollow tube.

As regards the intensity of the sound, 89% of the pupils questioned (N=28) answer correctly that the guided sound is "louder" at the end point. Nearly all pupils provide justifications, for example, that "The sound is trapped", or "There is no loss of sound", and particularly that, in the tube, no obstacle impedes the sound. What is significant however, is that more than half of the pupils (54%) answer, wrongly, that the speed of propagation[6] is greater when the sound is guided, alluding to the gain in intensity that has been obtained:

> "John receives more sound. He hears it sooner, because the sound reaches him a little sooner thanks to the tube that channels the waves, and therefore prevents loss."

> "The intensity of the sound is not the same, it is greater, because in this situation nothing disturbs the propagation of sound, since the sound is isolated. John begins to hear Peter first, because nothing slows down or disturbs the passage of sound to John."

When correlated, the answers obtained for both questions (boxes 2 and 3) suggest that the amplitude and the speed of propagation of the sound are linked, although the link is often implicit.

[6] Here the answers are not interpreted according to the distinction that physicists make between "group velocity" and "phase velocity" (see Moreau (1992) for an analysis of the problem and the experimental data). The pupils do not know of this distinction, and it would not, in any case, legitimate the association of "speed" with "amplitude."

Box 3

Guided sound: more intense, therefore faster?

Summary of the question

Peter is speaking to John. The only difference between situations 1 and 2 is that, in situation 2, Peter is speaking through a long steel tube.

Explain

1) whether, in situation 2, the intensity of the sound is greater, smaller, or identical to the intensity of the sound in situation 1. Why?
2) whether, in situation 2, John begins to hear Peter after the same amount of time as in situation 1, or sooner, or later. Why?

Results

For grade 11 science section students[7] (N=28) before instruction on propagation

- question 1, on the intensity of the sound:

different intensities[a]	**identical intensities**
89%	11%

- question 2, on the speeds of propagation:

different speeds[b]	**identical speeds**
79% V2>V1: 54% V2<V1: 25%	11%

- Pupils who answered "John hears the sound sooner in situation 2" (54%) used arguments such as "John receives more sound" and "The tube prevents sound loss", as if "**louder**" meant "**faster**."

a. Correct answer

b. In physical theory, several quantities are defined: phase velocity, group velocity (associated with energy), and wave speed (a characteristic of the medium). The correct answer, based on the evaluation of the "group velocity," is that John hears the sound later in situation 2.

4.3 Speed of sound during propagation

The results presented above recall those obtained for the rope questionnaire. In the pupils' arguments, the speed of propagation often depends on the source and amplitude of the signal. Once again, a hybrid concept in the pupils' reasoning may explain these associations. The "dynamic capital" supplied is a blend of "energy," "force," and "speed."

[7] Première scientifique (France).

Box 4

Does a sound that is getting fainter slow down?

Outline of the question

P• M• J•

Mary is at an equal distance from Peter and John on a straight road. Peter sings out an E. Does the sound take the same time to travel from Peter to Mary as it does to travel from Mary to John?

❑ YES Why?

❑ NO Why?

Results

At grade 9 (N=25), before any instruction on propagation, the times of propagation over the two distances are said to be

- identical (correct answer, i.e., YES): 48%
- different (NO): 32%
  "Because the sound gets fainter and fainter and therefore travels more slowly."
- no answer: 20%

If the "dynamic capital" changes during propagation (for example, if the medium is three dimensional), the pupils say that the amplitude and the speed of the sound both change simultaneously. According to them, a sound that is getting fainter slows down (box 4).

Specifically questioned as to whether the time of propagation of a sound over the two halves of a distance is identical or not, one third of the pupils answer that it is not, sometimes adding very explicit comments:

> "Because the sound gets fainter and fainter and therefore travels more slowly, like an earthquake."

4.4 The role of the medium

According to accepted physical theory, the decisive factor in determining the speed of a vibration is the medium. But from what has just been said, we can predict that, when the pupils take the medium into consideration, it is often as a passive support for the disturbance. If the latter is imagined as a moving object, then the medium may appear rather as an impediment. In view of this, do the pupils understand that sound cannot propagate in a vacuum? This is what the questionnaire summarised in box 5 seeks to determine.

Box 5
Is it possible to hear on the Moon?

Summary of the question

If a disaster occurred on the Moon (for example, an earthquake), would an astronaut orbiting around the Moon hear it? Why? Could we hear it on Earth? Why?

Correct answer

There is no air around the Moon; therefore the astronaut could not hear the earthquake (and neither could we, here on Earth!).

Results

Pupils in grade 9 (N=62), before any instruction on propagation:
Concerning the astronaut
- 40%: "The astronaut could not hear it": because (13%) "without air, there is no sound."
- 35%: "The astronaut could hear it."
- 25%: No answer.

Concerning a person on Earth
- 65%: "We couldn't hear it."
- 6%: "We could hear it."
- 29%: No answer.

Although a limited number of pupils think that an astronaut in orbit could not hear the sound of a disaster occurring on the Moon, only a third of the pupils (i.e., 13% of the 62 students questioned) associate this answer to the fact that there is no air. Most often, distance is cited as a decisive factor, with many who believe that the disaster could be heard in the vicinity of the Moon saying that it could not be heard on Earth, which is too far away:

> "We wouldn't hear it on Earth because the distance between the Moon and the Earth is too great. The astronaut might hear it, but at what distance is he orbiting the Moon?"

A final questionnaire bears more directly on the role of the medium. The pupils are asked to compare the propagation of sounds emitted by identical sources in different mediums: in air at different pressures and temperatures, in gaseous hydrogen, in water, and in a vacuum. Three quarters of them (N=39) say that sound cannot propagate in a vacuum, which is correct. But many also refuse the idea that sound can propagate in a liquid or in a solid: the denser the medium, the more it impedes propagation.

> "Yes, for a vacuum: nothing gets in the way of the sound."

"Yes, except for compact steel, because sound cannot pass through the filled tube to be recorded, and for water."

"Yes, for air, a vacuum, and hydrogen, because it doesn't act as an insulator like the water in tube 5."

"(Yes) except for steel, because it is a compact metal that does not let any air or water through, and therefore sounds will not be able to pass through it."

The answers obtained for the question on propagation speed confirm that many pupils believe that the denser the medium, the slower the propagation. Though they perceive that speed depends on the medium, the categories they establish do not correspond to reality; water and steel are more often seen as slowing down propagation than a vacuum, for instance:

"Some materials slow down propagation. From fastest to slowest, you have air, a vacuum, hydrogen, water, and steel."

Moreover, some comments elicited by another version of this questionnaire show that recognising that sound can propagate in a solid or a liquid is not enough to put an end to the preferential association between sound and the existence of a gas:

"Water: yes, there is oxygen in water."

5. CONCLUSION

There are marked resemblances between the responses concerning the propagation of a signal on a rope and those concerning the propagation of sound. Both phenomena give rise to reasoning that is based on a single notion, that of an object, such as was observed in connection with the dynamics of solids. This type of reasoning is, therefore, not simply due to the visual characteristics of the signal on the rope. It stems from very general trends of thought, such as the causal linear reasoning described in chapter 5. For the phenomenon of propagation, a simple, previous cause has to be found, and that is the source of the signal. The idea that some of this cause is supplied to the moving shape does the rest.

Based on probing questions and the identification of related difficulties, this research on the topic of sound leads to specific objectives and proposals for teaching. It suggests that very explicit comparisons should be made between the various types of "mechanics," such as the mechanics of solids and of mechanical signals (Maurines, 1988). It also reinforces the idea that it is not enough to say what a quantity depends on: one must also highlight surprising cases of independence (see chapter 9).

In addition, teaching the propagation of sound should be seen as one more opportunity to illustrate the limits of a form of reasoning that is so accessible and familiar to us – causal linear reasoning – and, in this way, to give greater significance to the descriptions proposed in physics.

REFERENCES

Maurines, L. 1986. *Premières notions sur la propagation des signaux mécaniques: étude des difficultés des étudiants*. Thesis. Université Paris 7.

Maurines, L. 1993. Mécanique spontanée du son. *Trema*. IUFM de Montpellier, pp 77-91.

Maurines, L. and Saltiel, E. 1988a. Mécanique spontanée du signal. *Bulletin de l'Union des Physiciens*, 707, pp 1023-1041.

Moreau, R. 1992. Propagation guidée des ondes acoustiques dans l'air. *Bulletin de l'Union des Physiciens*, 742, pp 1385-398.

Chapter 9

Constants and functional reduction

In association with Sylvie Rozier. This chapter is based on two studies, one, on constants, by the author (1982b), and the other on functions of several variables, by S. Rozier (1983). It makes use of material from an article published in *ASTER* (Viennot, 1992).

1. INTRODUCTION

Chapter 5 develops the following idea: common reasoning approaches the analysis of complex systems in a reductionist fashion. In quasistatic analysis, the multiple variables that determine the state of the system at a given time evolve simultaneously under the permanent constraint of relationships. In common reasoning, on the other hand, they are envisaged successively, as elements in a linear causal chain, evolving in a temporal succession. The principal aspect of this structure of reasoning is that the variables are studied one at a time.

One might find it surprising that students should be so reluctant to manipulate several variables, even at the university level. And yet, the problems studied in physics at secondary school almost always involve relationships in which there are several quantities. From the moment they begin to use algebraic formalism, pupils are taught to deal with such problems. The question is, what type of intellectual activity takes place in these cases?

Two points of view must be clearly distinguished. One approach is to use such relationships in an algebraic calculation, in order, eventually, to calculate a numerical value when several others are known. Another is to reason along the following lines: If such and such a quantity increases, and this other one is kept constant, then that one there will decrease. The first is the numerical approach, the second the functional approach.

The full importance of the functional aspect must be stressed. One might say that true understanding in any field, and especially in physics, comes with the mastery of functional dependences. It is, for instance, essential in checking the results obtained at the end of calculations.[1] Let us say a student considering the trajectory with radius of curvature R of a particle with mass m, charge q and speed v, in a magnetic field **B**, has inadvertently written the relationship R=qB/mv. If he/she re-examines this relationship in terms of functional dependences, noting that the radius of curvature obtained decreases when the mass and speed of the particle increases, and increases with the charge and magnetic field terms that are linked to the cause of the deviation, he/she will be able to realise that this result is incorrect (the correct relationship is R=mv/qB).

In secondary education, the numerical approach is given preference over the functional. In mathematics, the pupils manipulate single-variable functions; in physics, they use relationships involving two or more quantities, but essentially as a means of calculation. The idea of a functional dependence between several variables is not developed.

Children are already naturally inclined to apply reductionist methods in these situations. Thus, when commenting on a relationship such as the one linking the distance covered, the speed and the duration of travel, they often say: "Faster, therefore farther", or "Faster, therefore in less time" – inactivating, or rather ignoring, the third variable (Bovet et al., 1967; Crépault,1981).

As is amply shown in the first part of this book, students, too, are very often reductionist in their functional analyses. The studies summarised in this chapter highlight the particular characteristics of that reductionism in connection with a subject that may seem paradoxical here: "independence".

2. NUMERICAL OR FUNCTIONAL CONSTANTS

There appears to be nothing very complicated about the notion of "constants". And yet the term can have two radically different meanings:

- a numerical meaning, in which the noun "constant" is synonymous with a number that it is more or less useful to know, ranging from simple characteristics of objects such as the mass of the Earth, to what are

[1] For a study on result checking, see Serrero (1987).

known as universal constants, such as the Planck constant, h, or the speed of light in a vacuum, c;

- a "functional" meaning, in which the adjective has lost the noun it qualifies – a constant *function* of given variables. Functional statements are only meaningful if one has ascertained what variables the "constant" is *not* affected by, contrary to what might have been expected.

When we try to specify the variables that have no effect on the quantity being considered, we generally realise that this quantity depends on other variables. A functional, explicit, two-part approach can then be adopted, the first part bearing on "interesting independences," and the second on dependences. Box 1 gives an idea of how such an explicit approach can help make formulations that are commonly used in physics more precise and complete.

Box 1
Two common statements made explicit and complete

PROPOSED STATEMENT
The speed of light is a constant

Variables that have an effect on the quantity considered

The speed of light in a vacuum **does not depend...**	If not in a vacuum, the speed of light **depends...**
on the frequency	on the medium
on the frame of reference	on the frequency

PROPOSED STATEMENT
Ohm's law: at constant temperature, the resistance of a metallic conductor is a constant

Variables that have an effect on the quantity considered

When the type of conductor, its size and temperature are fixed, its resistance **does not depend...**	The resistance of a metallic conductor **depends...**
on the potential difference U across it	on the material of which the conductor is made
on the current I through it	on its dimensions (length and cross-section area)
	on its temperature

This research seeks to determine how students interpret statements that are heavily loaded with implicit connotations. When there is a choice of possible meanings, what are their preferences and questions? A survey conducted on this subject (Viennot 1982) among Science students in the first and second university years used the statements contained in box 1. The students are asked questions such as:

- In your opinion, is this statement clear and unambiguous?

- Does it seem incomplete? If so, what details do you think are necessary, or useful?

- Would you like to rephrase this statement? If so, how?

The most salient aspects of the results are outlined in tables 1 and 2.

Table 1
The speed of light is a constant

Quantity considered		Gembloux, 1st university year n=32	Gembloux, 2nd university year n=35	Paris 7, 1st university year n=35	Engineering school, 1st year n=100
does not depend	(1)	28%	6%	18%	9%
on the frame of reference	(2)	0%	6%	15%	3%
depends	(1)	50%	74%	85%	77%
on the medium	(3)	50%	68%	78%	77%

(1) Percentage of answers mentioning a non-dependence or a dependence, regardless of the argument.
(2) Percentage of answers mentioning a non-dependence with respect to the frame of reference.
(3) Percentage of answers mentioning a dependence with respect to the medium.
Gembloux: Agronomical Science College (Belgium).
Engineering school: Ecole Supérieure d'Informatique, d'Electronique et d'Automatique (Paris).

Table 2
At constant temperature, the resistance of a metallic conductor is a constant

The quantity considered		first university year (Belgium) n=32
does not depend on...	(1)	6%
depends on...	(1)	47%

(1) Percentage of answers mentioning a non-dependence or a dependence, regardless of the argument.
Gembloux: Agronomical Science College (Belgium).

The main thing to be derived from them is that, of the two constants considered, the speed of light and the resistance of the ohmic conductor, neither is reduced, at first, to a pure and simple number, such as c=300000km/s, for example. The functional aspect is envisaged, but, paradoxically, more often from the point of view of dependences than of independences. It is widely stressed that the speed of light depends on the medium. What might this quantity *not* depend on, in that case – i.e., how is it more "constant" than any other physical quantity? Very few students can say, and not one is worried about not knowing the answer. Studies on the resistance of a conductor which obeys Ohm's law bring similar results to light: only one student (out of the 41 that were questioned) spontaneously referred to the essential property of invariance, at a fixed temperature, with respect to the applied potential difference and the current through it, and two other students mentioned invariance in time; but the factors on which the "constant" depended were given in detail.

A similar reticence in making non-dependences explicit is to be found in textbooks and among teachers. Who, for example, thinks of specifying that the speed of a mechanical wave on a rope does not depend on how violently the rope was moved to begin with? And yet Maurines' study on this point (Maurines, 1986; Maurines and Saltiel, 1988a; see also chapter 4) shows how useful this would be. We think we've said it all when we've said that a quantity is constant, and that all that needs to be considered is what the constant might depend on.

If we take our analysis further, to a subtler and probably more conjectural level, and ask ourselves how these dependences are perceived and expressed, other points can be made.

One might expect that expressions such as "this quantity depends on that quantity", etc., would come most naturally. And yet what one very often observes are expressions resembling statement 2 (box 1):

> "when such and such a quantity is constant… then such and such another quantity is a constant;"
>
> "for a given medium… the speed of light is a constant;"
>
> "at a given temperature… the resistance… is constant."

This fact is probably not irrelevant, any more than is the infrequency with which the second statement in box 1 is reformulated as "the resistance… depends on the temperature" (17%). The two types of expression are not equivalent. It is very likely that the preference for the form "when X is constant, Y is constant" is due to the privileged role of time as an implicit variable in what we call constant functions, which leads to the interpretation summed up in the following diagram:

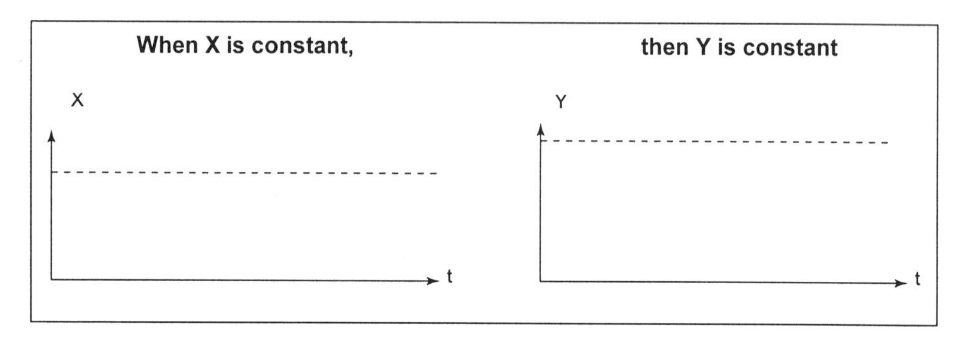

This interpretation brings the notion of a constant closer to that of a characteristic of an object, an object being defined by its permanence in time. It sheds light on concerns about possible dependences, such as those shown in the two comments below:

> "If the physical, climatic, and chemical conditions are constant, the resistance of an ohmic conductor is constant."
>
> "Conductors conductor which obey Ohm's law are rare. External variations other than temperature have to be taken into account. At a constant temperature, and at a given time, under given external conditions, the resistance of a metallic conductor is a constant."

It is tempting to sum up these comments as follows: if *everything* is constant, the resistance is constant. Of course, this very general summary drains the original statement of the meaning it was intended to convey (a non-dependence), but it does express another meaning, the one that the student gives it: that an object must be described in complete detail, for its characteristics to be clearly defined. For him/her, the constant is no more than a number on a tag that one sticks on an object.

This is probably why constants are so often introduced in connection with very specific characteristics of very specific objects, such as the mass of the Earth or of the Moon. Such a view of constant quantities clearly privileges the numerical, rather than the functional, approach. It is clearly linked to the tendency described in chapter 2, in which reasoning is anchored to the materiality of objects.

3. THE DIFFICULTY OF EXPRESSING NON-DEPENDENCES

The first conclusion to be derived from this is that non-dependences are not readily described as such. For any physical quantity, the list of "non-dependences" is endless. When studying constants, the whole point – and, sometimes, the main difficulty – is to determine those that are worth knowing.

Yet there may be an additional difficulty in expressing a non-dependence, as some variables in the problem under consideration may be linked to one another, that is, constrained by a relationship.

The common expression, "Such and such a quantity, G, does not depend on such and such another quantity, X", encourages a mistaken analogy with a mechanical device: pull on lever X, and G "moves," or "doesn't move". Not touching all the levers at the same time seems no more than sheer common sense. But, of course, if the variables describing the state of the system are mutually dependent, i.e., if the levers are connected to one another, complications arise. It is necessary to know how to define a set of independent variables from among these "state variables", which make it possible to define the system completely while respecting the incompatibilities due to the above-mentioned constraints. And then, when introducing any dependences of other quantities with respect to these variables, we would have to specify what happens to *all* these independent variables.

Thus, in physics, it is not enough simply to formulate Joule's law as is so often done: "The internal energy (U) of a given mass of perfect gas (i.e., where the variables pressure p, volume V, temperature T, and number of particles N are linked by the relationship $pV=NkT$, k being the Boltzmann constant) is independent of its volume". One must add "at constant temperature," since the internal energy in question depends on the product NkT ($U=3/2NkT$). One might equally well say, referring to the set of independent variables N, p, T, "The internal energy of a given mass of perfect gas does not depend on its pressure at a given temperature." Absurd juxtapositions might then seem valid:

"U does not depend on V, nor does U depend on p, therefore U does not depend on the product pV" (whereas $U=3/2pV$).

Non-dependence cannot be expressed simply, and the only simple formulation that is acceptable from the point of view of both common sense and physics is something of this sort: "G depends only on X, Y, Z...", but such a formulation does not mention the relevant independence – or independences.

Expressing these independences without ambiguity can take either the mathematical form of partial derivatives $\delta G/\delta A|_{x,y,z}=0$, or its verbal equivalent; the problem is that such a form is difficult to put into words, and the verbal translation is long ("The partial derivative of G with respect to A, if X,Y,Z are constant, is zero", or "G is independent of A, to the first order, when X, Y, Z... are constant"). Moreover, it makes it necessary to consider several variables simultaneously, which is probably the greatest obstacle. That, no doubt, is why so many oral statements, and even written texts, contain expressions that are incomplete, such as: "U is independent of V." The condition, "at a given T", is implicit.

A study conducted by Rozier (1983) sought to determine how the decoding process works. In individual interviews, some twenty French teachers (university or teaching institute[2] graduates with the highest qualifications[3]) were asked to react to the following passage on Joule's law, taken from a textbook :

> "Therefore, the internal energy of a given mass of perfect gas is independent of its volume."

The teachers are asked:

> "In your opinion, is this statement clear and unambiguous? Is there anything that you would add, or put differently?"

Their responses seem to fall into two categories:

- The specialist's reading (45%), where the phrase "independent of" is translated into terms from the formal register, for example, "$\delta U/\delta V=0$"; "V does not figure in the expression of U"; "dU=a dT"... Hardly any of the teachers in this group noted that the verbal statement was incomplete. Evidently, their analysis of such a statement is not based on what is said, but on the formal mechanism that is triggered.
- The common-sense reading (55%), where the text is taken to mean what it means to most people: no change in volume affects the internal energy. This statement is seen as ambiguous, and is completed with "at a given T" in order to make it acceptable.

It would seem, then, that the critical faculties are keenest among those teachers for whom words keep their common meaning; the other teachers are so used to the mechanisms that apply in this field that they do not realise that the statement is incomplete, and even incorrect.

[2] Graduates of the Ecole Normale Supérieure, Fontenay aux Roses.
[3] i.e., the agrégation.

It is interesting to see that even specialists apply a common-sense reading when they are faced with a term-by-term paraphrase of the statement, this time in connection with an everyday situation[4]:

	Statement	Implicit part of the message
Paraphrase	The price of a piece of carpeting does not depend on its length (a=length, b=width)	for a given surface (a.b=constant)
Quote from the text	The internal energy of an ideal gas is independent of its volume	at a given temperature (p.V=constant)

This time, the almost unanimous reaction is one of disapproval.

This study lays emphasis on the fact that the phrase "independent of" can lead to two types of interpretation – the first is formal, the second follows common lines of reasoning. Unless one is aware of this, one may find oneself, like the "specialists" mentioned above, incapable of understanding that others may not understand.

4. CONCLUSION

Again, it is evident that the idea of a material object prevails in common thought. It reduces the content assigned to concepts in accepted theory. Indeed, the whole point of constants, in physics, is that they go beyond the mere temporal permanence of a characteristic of an object. It is the exact specifications of invariance that give constants their full flavour, so to speak. Not to indicate what these specifications are is to sap the corresponding statements of their meaning, and to turn them into empty refrains.

What should be done in teaching? To begin with, one can apply the solutions suggested in the previous chapters: greater awareness on the part of both teachers and students as regards the difficulties associated with a subject, guidance in the planning of teaching activities, and, above all, well-defined teaching objectives.

[4] It may also be that a non-dependence is all the more difficult to apprehend as it cannot be related to the temporal permanence of an object. A given mass of gas, at a fixed temperature, does resemble an object, one whose "state" is all that changes (the terms of the invariant product pV). And that is probably why only one quantity – mass – is mentioned, although it has nothing to do with Joule's law, instead of the number of particles (N), which is the variable that needs to be specified here. It is more difficult to associate an invariance with a width/length product, the different forms of which (pairs of values of width/length) are not naturally associated with a single object.

Of course, taking into account the general aspects of reasoning presented here (see also chapters 5 and 6) is an added difficulty, pedagogically speaking. To what topic shall we devote the time that is necessary in order to make the rules of reasoning on multiple variables explicit, and to work on them? When can we help pupils develop the capacity to consider a result from the functional angle, rather than from just the numerical angle? (See also Saltiel, 1989; Maurines, 1991)

This implies an explicit and long-term determination. Our school textbooks, however, deal mostly with specific contents, and an objective in terms of a reasoning capacity may seem discouraging at first, and its effectiveness too diffuse. But those who decide to adopt these objectives do have lines of action open to them, even with the most elementary content-matter: it is possible to start working on multifunctional dependences as soon as the pupils have learned the area of a rectangle. A great deal is at stake.

REFERENCES

Bovet, M., Greco, P., Papert, S. and Voyat, G. 1967. Perception et notion du temps, Etudes d'épistémologie génétique, vol XXI, Paris, P.U.F.

Crepault, J. 1981. Etude longitudinale des inférences cinématiques chez le préadolescent et l'adolescent: évolution et régression, *Canadian Journal of Psychology*. 35,3

Maurines, L. 1986. *Premières notions sur la propagation des signaux mécaniques: étude des difficultés des étudiants*. Thesis. Université Paris 7.

Maurines, L. 1991. Raisonnement spontané sur la propagation des signaux: aspect fonctionnel, *Bulletin de l'Union des Physiciens*. n° 733, pp 669-677

Rozier, S. 1983. *L'implicite en physique: les étudiants et les fonctions de plusieurs variables*, Mémoire de D.E.A., Université Paris7, L.D.P.E.S.

Saltiel, E. 1989. Les exercices qualitatifs fonctionnels, *Actes du colloque sur Les Finalités des Enseignements Scientifiques*, Marseille, pp 113-121

Serrero, M. 1987. Critères de pertinence en physique, *Bulletin de l'Union des Physiciens*, n°699, pp 1229-1249.

Viennot, L. 1982b. L'implicite en physique: les étudiants et les constantes, *European Journal of Physics*, vol 3, pp174-180

Viennot, L. 1992. Raisonnement à plusieurs variables: tendances de la pensée commune , *Aster*, n° 14, pp 127-142.

Chapter 10

Rotation and translation: simultaneity?

In association with Jacqueline Menigaux. Based on a study by that author. The material presented here is taken from an article in *Bulletin de l'Union des Physiciens* (Menigaux, 1991); a few minor changes have been made.

1. INTRODUCTION

Considering the difficulties that students have with multi-variable problems, the motion of a rigid body may well be approached with trepidation: generally, no two points have the same motion. In fact, the linearity of the fundamental relationship of dynamics **F**=m**a** makes it possible to simplify the problem immensely, by characterising various aspects of motion: translation, rotation, and deformation. How do secondary school pupils and university students understand Newton's second law applied to a solid? More specifically, how do they envisage translation and rotation, two aspects of the motion of an object in space,[1] especially the fact that they are happening simultaneously? (See box 1).

The results of surveys on this topic can be interpreted in two ways: they bring to light components of common reasoning as well as characteristics of school learning.

[1] Deformation is not taken into account here.

Box 1

Newton's second law, translation and rotation

The most general motion of a solid that is free to move can be decomposed into simultaneous translation and rotation. To define them, it is best to start with the centre of mass G of the solid.

The motion of the centre of mass (G) of a solid is often said to be the motion of "translation" of the solid (even if this solid also rotates around G). Applying Newton's second law for a single particle (or "point mass") generally to all of the particles that make up the solid gives the following fundamental relationship (in a Galilean frame of reference):

$$\Sigma \mathbf{F} = m_T \mathbf{a}_G \tag{1}$$

where $\Sigma \mathbf{F}$ is the resultant of the forces applied to the solid,
m_T is the total mass of the solid,
and $\mathbf{a}_G$ is the acceleration of the centre of mass.

Although it describes the acceleration of the translation of the solid under the effect of the forces that are applied to it, relationship (1) is obviously identical to Newton's second law applied to a material point, point G, with mass m_T equal to that of the solid and submitted to the resultant of forces exerted on the solid. At this level of analysis, it is not necessary to know the exact point of application of the forces on the solid, it is enough to attribute to the solid under study the forces exerted on any one of its points.

The rotation of the solid is studied using a frame of reference with its origin at G, be this frame Galilean or not; here it is necessary to identify all of the forces and the precise point of application of each force. The theorem that applies concerns the change with time of angular momentum σ_G[a] about an axis through G:

$$\frac{d\sigma_G}{dt} = \sum \mathbf{M}_G \tag{2}$$

where $\Sigma \mathbf{M}_G$ is the sum of the moments about G of the forces acting on the solid.[b]

Relationships (1) and (2) are valid at all times as long as the speeds are negligible in comparison to the speed of light.

The motion of a rigid body generally involves simultaneous translation and rotation. If the system can be deformed, the deformation must be included in this motion. But in any case, relationships (1) and (2) remain valid.

a. The angular momentum about G of a particle (located at the point M, and with mass m and velocity $\mathbf{v}$) is the moment about G of the linear momentum vector $\mathbf{P} = m\mathbf{v}$ of that particle, or $\mathbf{GM} \wedge \mathbf{P}$. The angular momentum of the solid about G is the vector sum of the momenta of each of its particles.

b. The moment about G of a vector $\mathbf{A}$ acting through the point M is the vector $\mathbf{GM} \wedge \mathbf{A} = \mathbf{GM}.\ \mathbf{A} \sin\alpha\ \mathbf{k}$, where α is the angle between the vectors $\mathbf{GM}$ and $\mathbf{A}$, and $\mathbf{k}$ is a unit vector perpendicular to the plane of $\mathbf{GM}$ and $\mathbf{A}$.

2. THE INQUIRY: QUESTIONS AND RESULTS

A questionnaire was given to pupils in grades 11 (N=17) and 12 (N=18),[2] and to Science students in the first and second years of university (N=41, 42).[3] The following situation is presented:

A puck of radius R, initially at rest against one of the edges MN of a horizontal air table (a frictionless table) represented schematically by the rectangle MNPQ in the diagram in box 2, moves across this table under the influence of a constant horizontal force **F** (in the direction NP). A device ensures that the force always acts on the same point of the puck, either at point A (marked with a dot) or at point B (marked with a cross).

The participants were asked this question:

Will it take the same time, a shorter time or a longer time for the puck to hit the opposite edge PQ of the table when the constant force **F** is applied at point A than at point B?

The correct answer to this question, in view of the points recalled in box 1, is that the time taken by the puck to cross the table is the same in both cases, since it depends only on the translation of the puck, i.e., on the motion of the centre of mass G. And the motion of G remains the same when the point of application of the force changes (identical force applied, same initial and final positions, G_i and G_f, of the centre of mass). This answer is sufficient; however, one might add that, when the force is applied at B, the motion of the puck consists of a translation identical to that of the puck when the force is applied at A, and of a simultaneous pendulum-like rotation around the vertical axis passing through G.[4] In the absence of numerical data (mass and radius of the puck, intensity of the force, distance travelled), it is not possible to predict what point of the puck will hit the edge PQ.

Box 2 sums up the correct answer to the question and gives the answers obtained; according to more than 60% of all the participants, the puck takes

[2] Première scientifique (E) and Terminale scientifique (TC), the two last years of secondary school in France, science sections.

[3] DEUG A, France.

[4] The phenomena can be explained in terms of the work done by the force and the fact that the puck acquires kinetic energy as it moves. Thus, when the force is applied at B, it does more work than (or at least as much work as) the force applied at A since the distance of the projections on NP of the initial and final positions B_i and B_f of B is greater than (or at least equal to) A_iA_f (or G_iG_f). Therefore the final kinetic energy of the puck is greater (or at least the same) if the force is applied at B rather than at A. This means that when the puck hits the edge PQ, its final kinetic energy of translation is identical to that which it has when the force is applied at A, and its final kinetic energy of rotation is associated to its rotational motion. The value of the final rotational kinetic energy ranges from 0, when the projection on NP of B_iB_f is equal to G_iG_f, to a maximum value when the projection on NP of B_iB_f is equal to (G_iG_f +R).

longer to hit the edge PQ when the force **F** is applied at B than when it is applied at A.

Box 2
Why is it necessary to know the point of application of a force?

Situation

A puck on a horizontal table; a constant force **F** is applied always at the same point.

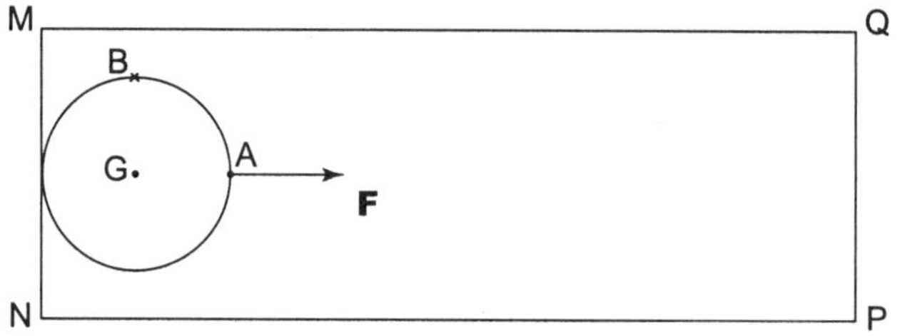

M Q B G. A F N P

Question

Will the puck take the same time, a shorter or a longer time to hit the edge PQ when the constant force **F** is applied always at point A than at point B?

Answers of pupils and students

	Same time	Longer time if force applied at B
Grade 11 (N=17)	6%	83%
Grade 12 (N=18)	22%	61%
First and second university years (N=53)	24%	67%

Correct answer

The same time whatever the point of application of **F**, because the time the puck takes to cross the table depends only on the motion of its centre of mass G, which is determined by the relationship $\mathbf{F} = m_T\mathbf{a}_G$.

More than 70% of the pupils and 20% of the students who answer that the puck will take longer if the force is applied at B give the following justification (see box 3): when the force F is applied at B, the puck accomplishes a rotation of $\pi/2$ before it translates along the direction of **F**, so that the line of action of the force goes through the centre of mass G; and only then is the translation motion of the puck the same as when the force **F** is applied directly at A, i.e., when the line of action of **F** passes through the centre of mass from the start:

> "The solid will first rotate by 90 degrees, then the solid will begin to move, so the total time (rotation time and translation time) is greater than the time of translation alone." (Grade 11)

"You have to wait for the moving object to rotate by 90 degrees for the force to be applied at A." (Grade 11)

"First, before moving, the object will rotate by a quarter turn so that B is in the place of A; after that, it has a uniform rectilinear motion."[5] (Grade 12)

"**F** will cause the solid to rotate on one spot for a limited time ($t <$ time of displacement) which will cause a delay." (Grade 12)

"The force begins by causing the puck to rotate; then the puck starts to move forward." (First year university)

"A rotation of $\pi/2$ before rectilinear displacement." (Second year university)

According to these explanations, the rotation of the object occurs prior to translation, until the solid is in a "suitable position" for translation. Although the two motions are produced simultaneously in the example chosen, the pupils and students dissociate them.

The simultaneity of rotation and translation is not, however, completely ignored in the answers given; 20% of the students and 6% of the pupils refer to it. But that does not mean that such answers are correct. Indeed, the puck is said to take longer to reach the opposite edge of the table when **F** is applied at B, because:

"For the same force, there will be a joint motion, a rotation around G and a translation." (Grade 11)

"The motion of the puck can be broken down into two parts: translation and rotation; this increases the duration of the displacement." (First university year)

"While moving forward, the solid will spin." (Second university year)

Translation and rotation are, apparently, not able to coexist without mutually affecting each other, and delaying the motion of the puck.

[5] More than one out of two pupils state that the puck has a uniform rectilinear motion. This error is still very widespread among the students. A common argument is that the resultant force exerted on the puck is zero – this is incorrect, but the belief is due to the use that is commonly made of the air table. Another error is that the force exerted on the puck is constant. This is probably due to a confusion between "uniform" and "uniformly accelerated," or to an incorrect and very frequent identification of force with speed (see chapter 3 and Viennot 1979, 1989a).

Thus, the point of application of the force on the puck is often wrongly considered to be a decisive factor in the duration of the motion involving translation and, possibly, rotation – whether or not they are understood as being simultaneous.

But at least, in the justifications quoted so far, the motion of the whole puck is analysed. This cannot be said for a third type of justification.

Box 3

Most frequent justifications

The puck will take longer to hit the edge PQ if the constant force **F** is applied at B because:

	The rotation of $\pi/2$ occurs before translation	The simultaneity of translation and rotation increases the time of crossing	Point B travels a greater distance than point A
Grade 11	73%	6%	13%
Grade 12	63%	0%	36%
First and second university years	20%	20%	37%

Box 3 shows that a considerable proportion of the pupils (36% in grade 12) and students (37% in the first and second university year) believe that the puck takes longer to cross the table when **F** is applied at B than when **F** is applied at A, basing their arguments on the fact that point B is more distant from the edge PQ than is point A.

"The distance ($x_{edge} - x_B$) is greater than ($x_{edge} - x_A$)." (Grade 11)

"B covers more distance, B is farther away than A." (Grade 12)

"Even though the speeds are the same, point B travels a greater distance than point A (points where the forces are applied)." (First university year)

"The force is applied over a greater distance." (Second university year)

As is the case for some comments mentioning the work done by the force **F** (rightly said to be greater when this force is applied at B), these arguments are most often (in 80% of the cases) based on a formula that is valid only for a point mass: $dW = \mathbf{F} . \mathbf{v}\, dt$.

The reasoning applied then only bears on a single point of the solid, which is not the centre of mass G, but the point of application of the force. Those who reason in this way are far from having understood that the translation of a solid does not depend on the point of application of the force, and, therefore, is independent of any rotation of the object about its centre of mass, all other things being equal.

3. DISCUSSION AND SUGGESTIONS

These results indicate that the majority of the pupils and students questioned apply at least one of the following aspects of reasoning in their approach to solids:

- a tendency to consider simultaneous motions as successive;
- a tendency to consider that the motions influence each other, even in those cases when the simultaneity of motions has been recognised;
- a tendency to analyse a single material point, that at which the force is applied.

The third tendency is probably largely due to school learning. At both the secondary school and university levels, teaching centres on the mechanics of a point mass. It is not surprising, given these conditions, that pupils and students should attach such a disproportionate importance to the exact point of application of force in the translation of a solid.

As for the first tendency, there may be two explanations.

In "causal linear reasoning," described in chapter 5, several concomitant phenomena are considered as successive.[6]

Moreover, what is learnt at school is likely to reinforce this tendency. Indeed, at both the secondary school and university levels, phenomena are studied separately – translation, then rotation, and finally, if necessary, deformation. The situations that are presented for study indicate that this is the preferred approach. As regards the translation of an object, its rotation is usually excluded by limiting the forces that it is subjected to or by considering it as "point-like"; it is presented as non-deformable. As for the rotation of an object (a non-deformable object, of course), global displacement is eliminated as the axis of rotation is specified as being fixed in space. Finally, the situation in which deformation is typically studied involves a spring, one end of which is fixed (so that no global displacement is possible) whilst the other end is subjected to a force along the axis of the spring (so that no rotation need be considered). Thus, pupils and students are accustomed to studying phenomena separately; although intended to simplify study, this separation easily leads to a belief that the phenomena considered are incompatible in time. It is therefore not surprising that some pupils and students should not understand the simultaneity of translation and rotation, or that they should exclude it altogether.

[6] See also Fauconnet (1981) and Rozier (1988).

Whatever it is that causes these forms of reasoning – whether a natural tendency that goes far beyond mechanics, or a consequence of teaching on this subject, or both – they are applied by more than two-thirds of the pupils questioned. The teaching community cannot ignore them. If we set ourselves the objective of making students understand the points mentioned here, how can we achieve it?

The main trends of natural reasoning cannot be modified by a single intervention (see chapters 5, 6, and 9). It is only by applying a series of measures, in the various fields of physics, that one can hope to redress the exclusive reasoning which considers "one variable" or "one phenomenon" at a time. Problems like the one in the questionnaire discussed here can be dealt with in this context.

For pupils to answer such questions correctly, would they need to know very much more than they are taught now? In secondary school and at university, the relationship $\Sigma\mathbf{F} = m_T\mathbf{a}_G$ is given, and often demonstrated. It is all that is needed to solve the problem presented above. However, this relationship is usually applied to a solid in translation, with a conveniently orthodox point of application of each force (a point that is "along the line of the motion" or through the centre of mass); such a choice restricts the scope of the relationship. The question described in this article is an example of a situation that can be proposed to make a formal relationship mean as much as possible. Finally, why is it that even though only the translation of objects is on the syllabus (for grade 12 science sections, established before 1994 in France), the exact point of application of forces is still made to seem so important? Arrows representing weight are carefully suspended from the centre of mass, and the reaction of an inclined plane on a block is represented at a precisely determined point of the side in contact with the plane. Those who still feel that this is absolutely necessary ought to consider that the end result may be what we have come across in this study: an exclusive centring on the point of application, when it is not relevant. Therefore, as always with this type of investigation, the main question raised is that of teaching objectives.

REFERENCES

Fauconnet, S. 1981. *Etude de résolution de problèmes: quelques problèmes de même structure en physique*, Thesis (Thèse de 3ème cycle), Université Paris 7.

Menigaux, J. 1991. Raisonnements des étudiants et des lycéens en mécanique du solide. *Bulletin de l'Union des Physiciens* n° 738, pp 1419-1429.

Rozier, S. 1988. *Le raisonnement linéaire causal en thermodynamique classique élémentaire.* Thesis, Université Paris 7.

Viennot, L. 1979. *Le raisonnement spontané en dynamique élémentaire*, Hermann, Paris.
Viennot, L. 1989a. Bilans des forces et loi des actions réciproques. Analyse des difficultés des élèves et enjeux didactiques, *Bulletin de l'Union des Physiciens*, 716, pp. 951-971.

Chapter 11

From electrostatics to electrodynamics: historical and present difficulties

In association with Abdelmadjid Benseghir and Jean Louis Closset. This chapter is based on a study by Benseghir (1989), and makes use of material from an article by Benseghir and Closset published in *Didaskalia* (1993).

The electric circuit is a familiar object nowadays, at least in schools. However, the difficulties that it poses are considerable.[1] Indeed, it requires an analysis in terms of systems and quasistationary change, and chapter 5 has shown how difficult this can prove. Here, the authors seek to explore the historical and didactical impact of an exclusively "electrostatic" view of the electric circuit, which is much more compatible with sequential reasoning than with systemic analysis.

Historically, electrodynamics appeared after electrostatics. And in teaching, too, it is usually broached after electrostatic interaction and electrical charges have been introduced. Thus, since previous knowledge influences new knowledge, it would not be surprising if the concepts and methodological approaches used in electrostatics constituted a determining, or limitating, framework for the elaboration of electrodynamics.

Has this been the case in the history of science, and is a similar process at work among pupils? It is true that historical progress and the development of personal knowledge do not always follow the same pattern. But our students might be coming up against the same problems as our forebears, as was seen

[1] See Tiberghien and Delacôte (1976), Johsua (1985), and Closset (1983, 1989).

in their approaches to changes in frames of reference and to impetus (chapters 3 and 4).

These questions are examined below, through an analysis of the principal stages in the history of the development of electrodynamics, and of the results of a survey conducted among French and Algerian students.

But first, let us go over a few elements of physics.

1. INCOMPLETENESS OF AN ANALYSIS CENTRED ON THE TERMINALS OF THE GENERATOR

The existence of charges at the terminals of the generator should not be used to explain the current in a circuit. Charges circulate inside the generator, too, or their accumulation would be dangerous. If the phenomena of attraction and repulsion of the electrons in the circuit by the charges at the terminals were to explain the current, that might do for external circulation, but not for internal circulation. Something else, then, must be capable of propelling the charges inside the generator *in spite of* the terminal charges. No matter what the mechanism behind it – electrochemical or inductive – this "something else" is called the "electromotive force." Sometimes the term "electromotive field" is used to designate the force per unit charge which corresponds to this action.

In an open circuit, the electromotive field makes the charges move towards the terminals until the electrostatic field resulting from this accumulation exactly counterbalances this electromotive field. In such a situation, therefore, there is a considerable asymmetry in charge at the terminals. But in a closed circuit, the circulation of the charges is not impeded by such an accumulation: the charges circulate in the entire circuit. There are charges at the terminals, too, but not only at the terminals: there are charges on the surfaces of all the components in the circuit. How else can one explain the fact that the electric field in the wires follows their shape exactly, no matter how tangled they may be? Again, this cannot be attributed solely to the charges at the terminals: the field of a dipole follows smoothly curved lines (called "field lines"), never those of a looped telephone cord, for example.

Until 1993, at least,[2] French textbooks were scarcely explicit as regards these major differences between open and closed circuits, between

[2] The 1993 syllabus for grade 8 (Quatrième) is explicitly aimed at clarifying the difference between open and closed circuits, as regards the roles of the charges located at the terminals of the battery (Quatrième is the third year of secondary schooling in France, and the first grade in which physics is taught as a full discipline).

concentrations of charge and a permanent current generator. A great many grade 8 textbooks, for example, stress the asymmetry of the charges at the terminals in a closed circuit (for instance, in France: Saison et al., 1979; Michaud and Lemoal, 1983; Averland et al., 1979, quoted in Benseghir, 1989). That might be taken to mean that asymmetry is responsible for the circulation of charges in the circuit. And what pupils are taught in the subsequent grades does not clear up this confusion.

At university level, as a rule, the change in the concentration of charges at the terminals is not discussed, nor is the existence of surface charge on the conducting wires.

The lack of clarity on this topic, which persists even in higher education, is not simply accidental. This clearly emerges from Benseghirs's study, which analyses two factors: the historical background and the reasoning students apply today.

2. HISTORY OF THE CONCEPT OF THE ELECTRIC CIRCUIT[3]

2.1 The science of electricity at the start of the nineteenth century

The "science of *electricity*," the predecessor of the electrostatics of today, was, until the end of the eighteenth century, the only approach to electricity. It dealt with two types of phenomena: attractions and repulsions between electrified bodies on the one hand, and electrical discharges and their various effects (sparks, shocks, etc.) on the other.

The electric properties observed were then considered to be the manifestations of a specific substance which existed in electrified bodies, "the electrical fluid" – subtle, weightless, incompressible and possessing a hypothetical elasticity and expandability which was termed "expansive virtue or force". The patent substantialism of that time (Bauer, 1948) is to be found among students, too.

According to the two fluid theory, which was especially strong in France, two kinds of electricity are defined with respect to a certain "natural state" of bodies, described as follows:

> We know that all bodies contain a certain quantity of electrical fluid; that this fluid can be considered as being composed of two different fluids;

[3] What follows is taken from the article by Benseghir and Closset cited at the start of this chapter.

> namely: vitreous or positive fluid, and resinous or negative fluid. That as long as these two fluids which constitute the electric fluid are combined, they do not manifest themselves in any way. But as soon as, through some circumstance one or the other of these fluids, or both, become free, they give the bodies within which or on the surface of which they reside, the property of attracting or repelling one another. (Thenard, 1813)

By the end of the eighteenth century, physicists held the "science of electricity" to be highly reliable. Haüy voiced this feeling in 1803:

> Electricity, enriched by the work of so many distinguished physicists, seemed to have reached the point where a science need make no further strides, leaving the next generation of scientists only the hope of one day confirming the discoveries of their predecessors and of shedding more light on the truths that had already been revealed.(Haüy, 1803)

It was in this climate that the first experiments in electro-dynamics were received and interpreted.

2.2 The transition from electrostatics to electrodynamics

Galvani was the first to make systematic use of situations involving closed circuits in his experiments, in 1789. The general principle was to put two different metals (silver and zinc, for example) in contact and to connect one to the muscle of an animal, and the other to the nerve of that muscle (figure 1). When this device, then called a "galvanic circle, or chain," was closed, the muscle would contract.

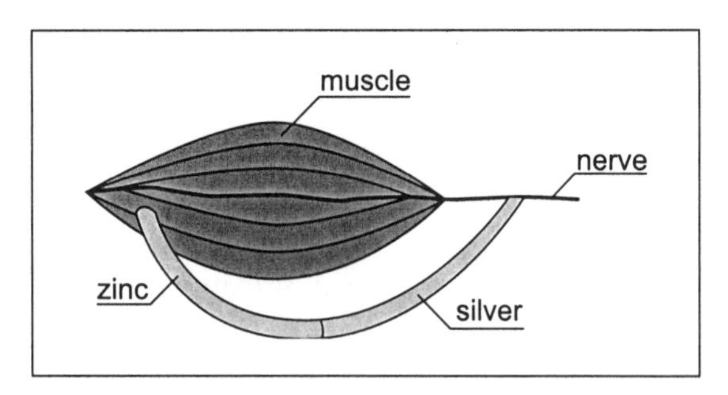

Figure 1 Galvanic circle

What could account for such phenomena? At the time, for lack of a more specific idea, this new development was explained by appealing to an indeterminate substance: the "galvanic fluid".

The first hypothesis on record is that of Galvani himself, and testifies to the importance of vitalist views at that time: the cause was to be found in "the nervous electrical fluid," a type of electricity that exists in organised bodies. Volta rejected the thesis of "animal electricity," maintaining as early as 1792 that the "galvanic fluid" was nothing but ordinary "electrical fluid" (Volta 1801a). This fluid, he said, came from the contact of the two different metals, and the electrical imbalance that Galvani spoke of occurred in the "conducting arc."

But Volta himself, faced with the difficulty of backing his theories with experimental data on the closed "galvanic circle," had to operate in an open circuit, and therefore brought the problem back into the realm of electrostatics (Polvani, 1949; Varney and Fischer, 1980). He described his experiments thus:

> Both metals (silver and zinc) are cleaned and polished; they are placed in immediate contact at one or several points; they displace the electrical fluid, break its equilibrium, so that the fluid passes from the silver to the zinc, is rarefied in one and condensed in the other, and is maintained there in that double state of rarefaction and condensation.... (Volta, 1801b)

In order to explain electrification by contact, Volta introduced an ill-defined quantity, called "the electrical motive force"; located on the surfaces where the two different metals were in contact, it caused the electrical imbalance between the metals. Towards the end of 1799, basing his ideas upon this "contact theory," Volta constructed the first "electromotor" by stacking identical pairs of different metals between pieces of damp cardboard, whose presumed role was simply to conduct the electrical fluid and to ensure the additive effect of the electricity generated in each pair.

Volta seems to have made a clear distinction between the electrostatic effects of the battery in an open circuit and the effects obtained in a closed circuit. He accounted for the latter by attributing to the "electrical motive force" the role of continuously propelling the electrical fluid in the "conducting circle." As he put it,

> ...metals and various other bodies are not merely that, or merely conductors, bodies that are permeable by the electrical fluid (...), but true exciters or perpetual motors of that fluid, so that when they are put in contact with each other, they form a continuous and returning chain, or, and this amounts to the same thing, they form a conducting circle which maintains a continuous electrical current that is not slowed down (Volta, 1802).

It would seem that he was describing a steady state.

2.3 The new reduced to the old

Some of the French physicists before whom Volta defended his thesis in 1801 had contributed greatly to the development of the "science of electricity." Loath to accept the dynamic aspect of his theory (Brown, 1961), they reduced it to the principle that electricity develops through contact in an open circuit. In a report written in 1801 for a commission of the *Institut*, Biot makes no mention at all of the idea of a "continuous current of electrical fluid". Haüy's comments on the report show his reservations about the idea of impulsion of the electrical fluid:

> That famous physicist (Volta) seems to accept that there is an impulse which drives into the zinc a part of the electrical fluid that belonged to the copper (...). We think it preferable to deal with it in the same way as with electricity produced by friction or by heat, that is to say, to limit ourselves to a strict account of the facts, without venturing to consider the force producing motion, about which there does not seem to be sufficient information. (Haüy, 1803b)

The tendency to reduce the new to the familiar designated the "science of electricity" as the appropriate framework in which to study the phenomena obtained with the voltaic pile. Haüy went on to write that:

> The results of the new investigations that must be carried out to dissipate the cloud that still hangs over this area of science will certainly not establish an essential distinction between galvanism and electricity, but only conciliate electricity to itself. (Haüy, 1803)

Parallels with "electrostatics" were established through an analogy with the Leyden jar, in other words, with a capacitor. Nearly fifty years later, Pouillet was to write:

> The Volta pile is, therefore, really a Leyden jar that recharges on its own and runs out only after a very long time, instead of running out after each discharge, like an ordinary battery. (Pouillet, 1847)

Thus, the phenomena involved were apprehended through conceptions of familiar phenomena, with some minimal modifications to integrate the new phenomenological aspects. Such a process is in accordance with a relatively general epistemological approach (Kuhn, 1970), which can also be observed in teaching (Johsua and Dupin, 1988).

Previous knowledge of "electrostatics" also left its mark on the methodology of the time. Working from earlier conceptions, physicists focused their research on the isolated battery, that is, on a battery in an open circuit (Brown, 1969; Blondel, 1982). What could be "seen" in the new phenomena was, therefore, only what one was prepared to "see." Thus, Pfaff was to write (1829):

> Moreover, it is in this state, before the two ends are joined to form the chain itself, that the phenomenon presents itself in its greatest simplicity and that one may hope to find the truly essential conditions, more surely than with a more complicated combination.

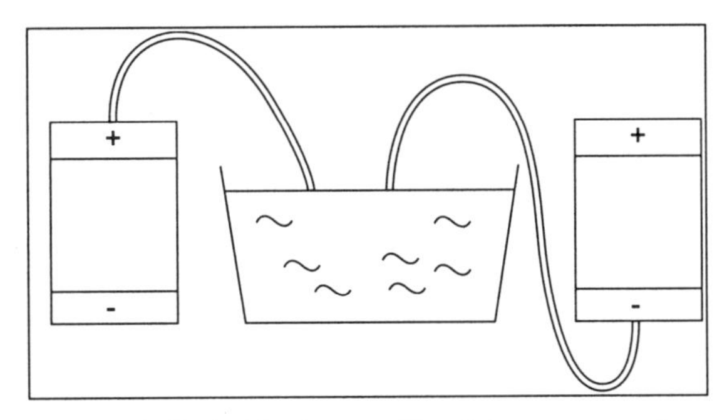

Figure 2 De la Rive's experimental arrangement

In addition to reproducing and studying "electroscopic" effects, scientists tried doggedly to produce chemical and magnetic effects from the terminals of an isolated battery. When the electrolysis of water was first achieved, efforts were made to produce gases by the direct action of a single terminal of the battery. Cuvier refers to this procedure (1801):

> They all began their research in the same way, to find out if it is possible to produce the two gases in separate waters. If the waters are absolutely isolated, no gases appear.

Although it had been established that such procedures were ineffective, attempts to observe them were made by famous physicists, as is shown in this passage from De La Rive (1825, illustrated in figure 2):

> Thus I ascertained that it is impossible to produce chemical decompositions by immersing only one end of the pile in the liquid; nor is it possible if the two opposite ends of two different piles are immersed.

The experiments conducted by Oersted in 1820 can be seen as initiating the second transition from static electricity to dynamic electricity. They show that a magnetic needle deviates when placed next to the conducting wire connecting the ends of a voltaic pile. Breaking with the experimental procedures of his contemporaries, Oersted clearly explained that the "circuit" had to be closed for this new phenomenon to occur. But he attributed the cause to the two terminals, referring to an "electric conflict" –

an expression which, in the eighteenth century, had been used exclusively for the effects of discharge between terminals (Bauer, 1948).

Although the works of Ampere from 1820 onwards showed that there were properties common to the battery and the connecting conductor (magnetic needles deviated in their vicinity), the electric current was at that time still commonly interpreted in terms of the model of antagonistic currents: the terminals of the battery were considered as "indefinite sources of contrary electricities" (Pouillet, 1828), while "the conductor between the terminals was a constant link between the accumulated electricities" (Lamé, 1937); no other mention was made of the possibility that there was current in the battery.

It was not until 1836 that something approaching a systemic view of the electric circuit was presented by Peltier:

> The battery and the connecting conductor constitute a single system, all the parts of which are connected, so that the electromotor is no longer under the same conditions when the conductor is modified, for example. (Peltier 1836)

3. THE REASONING OF STUDENTS TODAY

Is the systemic vision, which has been so laboriously elaborated in the course of history, that of students today? Or do they, in fact, apply elements of analysis that are much closer to electrostatics, such as supplies of charges and sudden discharges?

That is the subject of the following section.

The study conducted on this topic concerns French and Algerian university students and school pupils – see table 1 for details. Unless otherwise indicated, the participants were questioned after instruction on electrodynamics.

Table 1
Levels of education of the participants in the study

ALGERIA	FRANCE
- 2e AS (French Première, i.e. grade eleven) - 3e AS (French Terminale, i.e. grade twelve)	- Seconde (i.e. grade ten) - Première (i.e. grade eleven) - Terminale scientifique (i.e. grade twelve, Science section)
- Univ. 1 first university year, Science students	- DEUG 1 - DEUG 2 first and second university years, Science students

The first situation (fig. 3) used in the empirical study is directly inspired by De La Rive's experimental device (fig. 2).

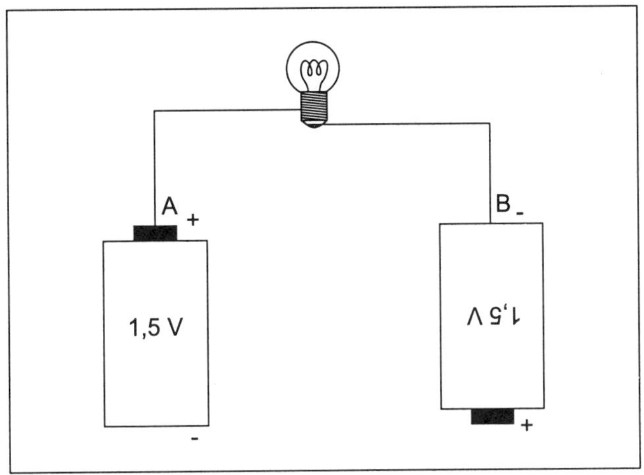

Figure 3 Situation 1

Question:
Is the bulb lit? YES NO Why?
Because the circuit is not closed, there is no potential difference between the terminals A and B and the bulb is not lit. The responses obtained are given in table 2.

Table 2
Percentage of answers obtained for situation 1

	FRANCE				ALGERIA	
	Seconde, prior to instruction	Seconde, after instruction	Première	DEUG 1	2ème AS	Univ. 1
YES	69	19	10	16	55	51
NO (correct)	31	81	90	82	44	49
No answer	0	0	0	2	1	0
N=	35	53	60	83	112	94

The correct answer (NO) is given by the great majority of French participants after instruction. Surprisingly, though, 16% of the DEUG students predict that the bulb will light up in this situation. The justifications for the correct answer are essentially based on the argument that the electric circuit is not closed, and are purely operative:

> "First, I tried it and the lamp did not light up, and in order for the lamp to light up the circuit must be closed, and in this case it is not." (2ème AS)

The incorrect answer (YES) was given by the great majority of Seconde pupils before instruction on the subject, and by the majority of Algerian students even after instruction.[4]

The justifications provided for the YES answers were almost all based on the difference in the signs of the charges at the terminals:

> "Because the electrons of battery 2 are negatively charged and are attracted by the + end of battery 1. Therefore, they will pass through the filament and light the bulb." (DEUG, 1st year)

> "Because the electric current normally goes from + to - . The circuit can be open because there are two generators." (Seconde)

This is definitely reminiscent of Pouillet's "indefinite sources of contrary electricities" at the ends of the battery.

One pupil even asserts that the bulb must light up in spite of evidence to the contrary, furnished by the experiments conducted beforehand:

> "Because whatever the two batteries (...), electrons will pass from one to the other. But I tried this experiment and it did not work." (2ème AS)

Such a comment perfectly illustrates the following remark by Greco and Piaget (1959):

> The truth of the idea is more coercive than that of the fact, if the idea does not stem from arbitrary conjecture but from a certain logic, be it correct or incorrect, or, to put it more accurately, complete or incomplete.[5]

As for the correct answers (NO), they cannot be seen as a sufficient guarantee that the concept of the electric circuit has really been mastered. Indeed, it is possible that once the need for the circuit to be closed has been established, it serves as a provisional defence against an erroneous answer. Situation 2 was devised to shed light on what may be instances of masked incomprehension (fig. 4). It was proposed to the pupils, along with the same question as before:

[4] In Algeria, electricity is taught only in the fifth year of secondary schooling, and electrodynamics is dealt with after substantial instruction in electrostatics. In the first university year, the same approach is generally adopted (classes in electrostatics followed by the analysis of electric circuits). The fact that more emphasis is laid on electrostatics probably reinforces the students' tendency to apply previous knowledge erroneously in this field.

[5] See also Johsua and Dupin, 1988.

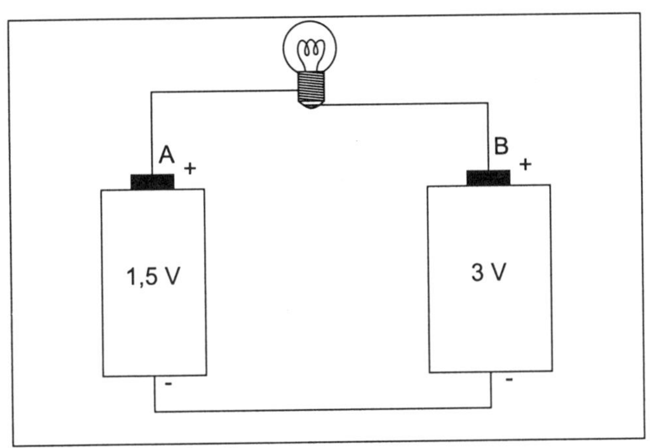

Figure 4 Situation 2

Is the bulb lit? YES NO Why?

The responses to the question are presented in table 3.

The differences in the results may easily be explained by the fact that the first class was questioned prior to instruction on the subject, and that the Première pupils were not from science sections, whereas the Terminale pupils were in a science section. The most salient fact is still, however, that, in all cases, a very large majority of pupils think that it is not possible for the bulb to light up. How do they justify this?

Table 3
Percentage of answers obtained for situation 2

	FRANCE				ALGER-IA		
	Seconde, prior to instruct-ion	Seconde, after instruct-ion	Première	Termin-ale	2ème AS	3ème AS	Univ. 1
YES (correct[a])	6	24	10	41	34	20	35
NO	91	76	90	57	64	77	56
No answer	3	0	0	2	0	3	9
N=	35	21	103	44	204	71	45

a. Provided the bulb has characteristics that match the situation.

"The bulb cannot light up, because it is between two positive ends, and therefore the current does not circulate." (Première)

"It cannot light up because there is a repulsion between the two ends, since they are both positive." (2ème AS)

"Because for a bulb to light up, a + current and a – current have to reach it, but in this diagram it receives two + currents." (Seconde, prior to instruction).

The last comment, in which traces of a model of antagonistic currents can be detected, is not fundamentally different from the others. They all make use of the "lack of asymmetry of charge at the terminals" argument, and centre unduly on the terminal charges, as if localised supplies of charges explained everything.

4. CONFUSIONS BETWEEN CHARGE AND POTENTIAL

Thus, the understanding of electric circuits is hindered by a sort of invasion of the idea of charge, which, in this case, is associated with the terminals of the battery. The potential difference then becomes a captive concept, with no real autonomy: if there is no difference in charges, or if there are no charges at all, there can be no potential difference. We might, therefore, predict that even within electrostatics, a similar reduction could be observed.

This is what situation 3 (fig. 5) seeks to determine.

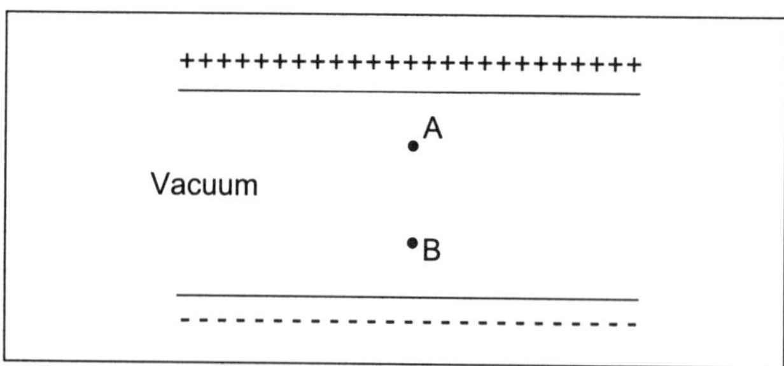

Figure 5 Situation 3

The participants are asked the following question:

A and B are two points situated in the space between two conducting plates, respectively positively and negatively charged, and facing each other in a vacuum. Is there a potential difference between points A and B?

YES NO Why?

Table 4 Percentage of answers obtained for situation 3

	FRANCE				ALGERIA	
	Seconde	Première	Terminale	DEUG 1	3ème AS	Univ. 1
YES (correct)	33	22	48	59	37	30
NO	67	64	52	32	63	65
No answer	0	14	0	9	0	5
N=	70	14	27	83	38	74

Table 4 presents the percentage of answers obtained. Apart from a single DEUG population, the great majority answered incorrectly, predicting that there would be no potential difference between A and B. This is usually justified by stating that there is no charge at these points, as the following comments show:

"Points A and B are neutral; they are not charged with opposite signs." (Seconde)

"A and B are not material. They have no charge, therefore there is no difference in potential between A and B." (DEUG 1)

"A and B are not charged with electricity; the two plates cannot have any effect on these two points." (2ème AS)

Thus, the highly abstract concept of potential appears here in a reduced form, and seems very close to the concept of electric charge. As electric charges are presented in the very first lessons on electricity, they are given a strong causal status from the outset. That is most probably why the presence of charges at one point has become a *sine qua non* condition for the existence of other quantities at that same point. The study on the electric field presented in chapter 12 supports this conjecture.

5. CONCLUSION

Definite correlations can be established between the two parts of this study.

The historical analysis shows that, for several decades, phenomena concerning the electric current were apprehended through operational conceptions and approaches that were related to electrostatics.

The study conducted among students today shows that many of them base their reasoning on a more or less intuitive knowledge of electrostatics. The reference frameworks adopted for the study of the electric circuit, historically and individually, are comparable, as can be seen from the similarities between historical approaches and students' reasoning today.

The most striking example of this similarity is the attitude that consists in focusing on the terminals of the battery.

In the historical context, situations involving closed circuits were only seldom considered in research and interpretation, as these were circumscribed by the reference framework of static electricity. New phenomena could only be seen as an extension of electrostatic phenomena or discharge phenomena (in the sense the term had then). So the distinction between open and closed circuits could not be seen as significant at that time. The context explains why the terminals were ascribed a crucial role.

Pupils reason along similar lines, focusing on the terminals that they "see" as "isolated." They attribute to these terminals charges that they consider to be the essential factor for the analysis of the circuit. Therefore, the distinction between open and closed circuits does not appear to be relevant.

From 1820 onwards, it was increasingly acknowledged that by closing the circuit, it was possible to obtain phenomena that could not be reduced to static electricity. The notion of a circuit that emerged then did not, however, correspond to any clear-cut idea of electric circulation. It only implied a morphological continuity of the devices constructed. Scientists focused on the terminals; then this was adapted to the new object, the closed circuit. The "excess electricity" that developed at the terminals of the battery was considered as generating an "electric current" in the "connecting" conductor – in fact, a combination of two antagonistic currents.

Even though most of the pupils questioned no longer apply the model of antagonistic currents (which is still found among younger children: see Tiberghien and Delacôte, 1976), their behaviour, too, is only partially adapted to the operational need for a closed circuit. The most frequently observed model corresponds to the motion of charges produced at one end of the generator, and drawn by the charge at the other terminal which has the opposite sign.

The parallel established in this study can no doubt be explained by a certain economy of thought: to apprehend the real, what is most accessible is the idea of a cause materialised by an object. Here, the charges serve that purpose.[6]

It is difficult to determine how, once the implications of these results have been perceived, one can hope to make pupils understand the difference between a steady current and a discharge in a circuit, without being carefully explicit.

It seems difficult to achieve this without prudently resorting to an analogy. At first sight, an analogy with the circulation of a fluid in a horizontal circuit (a transparent one, in this demonstration) has the advantage of adequately illustrating the circulation of the whole, including what takes place inside the pump which represents the generator. But it is not easy to take this analogy further, and to illustrate the differences in potential through differences in pressure – simply because the notion of pressure in a fluid is far from clear for pupils (Closset, 1992). Similarly, devices such as bicycle chains (Closset, 1983, 1989), i.e., mechanical and closed systems in which interacting elements move due to the effect of a localised motor, can illustrate both circulation without accumulation (comparable to that of charges) and the localisation of the motor (comparable to the generator). But again, the analogue of the potential difference (the difference in tension between two points of the chain) is not very accessible. Moreover, it is more difficult to illustrate what resistances in parallel are in this model than with fluids. The incompleteness of these analogies is in fact typical of all analogies. Drawing an analogy is a good opportunity to point out these limits.

Whatever the case, introducing the steady current generator by comparing it with the equivalent of a discharging capacitor constitutes a risk that must be recognised: that students may never realise that, among the elements which make physical analyses possible, there is a quantity known as the "electromotive force," which accounts for the circulation of charges where electrostatics alone cannot.

REFERENCES

Bauer, E. 1948. *L'électromagnétisme, hier et aujourd'hui. Paris,* Albin Michel.

Benseghir, A. 1989. *Transition électrostatique-électrocinétique : point de vue historique et analyse des difficultés des élèves.* Paris, Thèse, Université Paris 7.

[6] References to the article cited above end here.

Benseghir, A. and Closset, J.L. 1993. Prégnance de l'explication électrostatique dans la construction du concept de circuit électrique: points de vue historique et didactique. *Didaskalia* 2, pp 31-47.
Benseghir, A. and Closset, J.L. 1996. The electrostatic-electrokinetic transition: historical and educational difficulties, *International Journal of Science Education*, vol 18, n°2, pp 179-191.
Biot, J.B. 1801. Rapport fait à la classe des sciences mathématiques et physiques de l'institut national sur les expériences du cit. Volta. *A.C.*, n° 41, pp 3-23.
Blondel, C. 1982. *Ampère et la création de l'électrodynamique*. Paris, Bibliothèque Nationale.
Brown, T. 1969. The electric current in early nineteenth century French physics. In *Historical studies in the physical sciences*. vol. I, pp 61-103.
Closset, J.-L. 1983. *Le raisonnement séquentiel en électrocinétique*. Paris, Thesis, Université Paris 7.
Closset, J.-L. 1989. Les obstacles à l'apprentissage de l'électrocinétique. *B.U.P.,* n° 716, pp 931-950.
Cuvier, G. 1801. Rapport sur le galvanisme. *Journal de Physique*, n° 52, pp 318-324.
De La Rive A. 1825. Mémoire sur quelques-uns uns des phénomènes que présente l'électricité voltaïque dans son passage à travers les conducteurs liquides. *A.C.P.*, n° 28, pp 190-221.
Greco P. and Piaget, J. 1959. *Apprentissage et connaissance*. Paris, P.U.F.
Haüy R.J. 1803. *Traité élémentaire de physique*, tome 1, 1ère édition.
Johsua, S. 1985. *Contribution à la délimitation du contraint et du possible dans l'enseignement de la physique (essai de didactique expérimentale)*, Thèse d'état, Marseille, Université de Provence.
Johsua, S. and Dupin, J.J. 1988. La gestion des contradictions dans les processus de modélisation en physique, en situation de classe. In *Didactique et acquisition des connaissances scientifiques, Actes du colloque de Sèvres*, mai 1987. Paris, Edition de la pensée sauvage.
Jouguet, M. 1955. *Traité d'électricité théorique*, tome 2. Paris, Gauthiers-Villars.
Kuhn, T.S. 1970 *The structure of scientific revolutions*. Chicago, Il.: University of Chicago Press.
Lamé, G. 1837. *Cours de l'école polytechnique*, tome 2, 2ème partie.
Michaud, Y. and Lemoal, Y. 1979. *Physique 4ème*. Paris, Magnard.
Pfaff, C.H. 1829. Défense de la théorie de Volta relative à la production de l'électricité par le simple contact, contre les objections de M. le professeur De La Rive. *A.C.P.*, n° 41, pp 236-247.
Peltier, J.C.A. 1836. Courants électriques. Définition des expressions Quantité et Intensité. *Comptes rendus de l'Académie des sciences*, n°2, pp 475-476.
Polvani, G. 1949. L'invention de la pile. *Revue de l'histoire des sciences*, n° 2, pp. 340-351.
Pouillet, C.S.M. 1828. *Eléments de physique expérimentale*, tome 1, 1ère édition, p. 635.
Pouillet, C.S.M. 1847. *Eléments de physique expérimentale,* tome 1, 5ème édition, p. 596.
Rosser, W.G.V. 1970. Magnitudes of surface charge distributions associated with electric current flow. *Am. J. Phys.*, n° 38, pp 265-266.
Rozier, S. 1988. *Le raisonnement linéaire causal en thermodynamique classique élémentaire.* Paris, Thesis, Université Paris 7.
Sanner, M. 1983. *Du concept au fantasme*. Paris, P.U.F.
Saison A., Malleus, P., Huber, P. and Seyfried, B. 1979. *Physique 4ème*. Paris, Nathan
Thenard L.J. 1813. *Traité élémentaire de chimie,* tome 1, 1ère édition.
Tiberghien A. and Delacôte G. 1976. Manipulations et représentations de circuits électriques simples par des enfants de 7 à 12 ans. *Revue française de pédagogie*, n° 34, pp 32-44.

Varney, R.N. and Fischer, L.H. 1980. *Electromotive force : Volta's forgotten concept.* Am.J.Phys., n° 48, pp 405-408.

Volta, A. 1801a. Lettre du professeur Volta à J.C. de la Mètherie sur les phénomènes galvaniques. *Journal de Physique*, n° 53, 309-316.

Volta, A. 1801b. De l'électricité dite galvanique. *A.C.*, n° 40, pp 223-256.

Volta, A. 1802. Lettre de Volta sur l'identité du fluide électrique avec le prétendu fluide galvanique, à M. Bancks. *Journal de Chimie*, n° 2, pp 158-169.

Chapter 12

Superposition of electric fields and causality

In association with Sylvie Rainson. This chapter is based on a study by that author (Rainson, 1995), and makes use of material from articles by Viennot and Rainson (1992) and Rainson, Tranströmer and Viennot (1994). See also Viennot and Rainson (1999).

1. INTRODUCTION

The electric field is one topic in physics that might be expected to discourage common reasoning, as the theoretical concepts involved are very advanced. But, once again, it appears that some forms of reasoning are at work which no school would admit to have taught.

The physics content can be summed up as follows:

Let us consider a given number of electrically charged particles q_1, q_2... q_i (henceforth referred to simply as "charges"). For given positions, we can say that the interaction between each pair of charges is the same as if the charges considered were alone in space, and then add the forces calculated in that way to find the total force **F** exerted upon each of them: this is the superposition principle.

Thus, it is possible to describe the electrostatic interactions by using just one simple law: Coulomb's law (box 1). According to this law, the total force acting upon a charge q_i is proportional to that charge (if the rest of the situation is "fixed") because the strength of all the interactions in which it is involved are proportional to it. The situation can therefore be described in terms of an electric field: this quantity is defined at each point by the relationship $\mathbf{F} = q\mathbf{E}$.

Box 1

Coulomb's law and the superposition principle

Let q_1 and q_2 be two electric charges. They interact.
Charge q_1 exerts on charge q_2 a force given by Coulomb's law

$$\mathbf{F}_{1\,on2} = \underbrace{\frac{1}{4\pi\varepsilon_0}}_{\text{constant}} \frac{q_1 q_2}{r^2}\mathbf{u}_1$$

$\mathbf{u}_1$ is a unit vector situated as on the diagram
r is the distance between the charges

q_1 experiences a force $\mathbf{f}_{2\,on\,1} = -\mathbf{f}_{1\,on\,2}$ (Newton's Third Law).

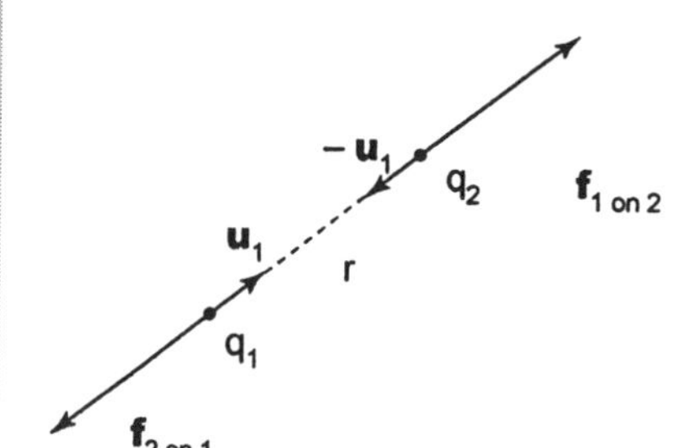

Illustration of the situations where the charges have the same sign

One can also say that charge q_1 is the source of an electric field in M2 (position of q_2)

$$\mathbf{E}_1(M_2) = \frac{q_1}{4\pi\varepsilon_0}\frac{\mathbf{u}_i}{r^2} \quad \text{such that} \quad \mathbf{f}_{1\,on\,2} = q_2\mathbf{E}_1(M_2)$$

If several charges act upon another charge q, placed at M, the total field at M, E(M), is defined by the equation F=q (M), where F is the force experienced by q at the point M.

According to the superposition principle, this field can be expressed as the sum of the contributions of each charge calculated as if each were the only one to interact with q.

$$\mathbf{E}(M) = \sum_i \mathbf{E}_i(M)$$

where $$\mathbf{E}_i(M) = \frac{1}{4\pi\varepsilon_0}\frac{q_i}{r_i^2}\mathbf{u}_i$$

q_1 q_2
$\mathbf{E}_i$
M q_i
$\mathbf{E}$ q_3

Being $\mathbf{E}_i(M)$ = contribution to charge q_i of the electric field in M;
r_i = distance between charge q_i to point M;
$\mathbf{u}_i$ = unit vector directed from q_i toward M.

Knowledge of the field makes it possible to predict what will happen to any charge which happens to be placed at the point considered, as a result of the positions and values of all the other charges.

In short, the superposition principle may be summed up thus[1]: every unit charge present at a given time in a quasistatic situation (see chapter 5) contributes in the same way to the electric field at any point M in space, according to Coulomb's law.

This principle may seem unsurprising and obvious, but is it easily accepted? Is it correctly applied to all the situations that the statement encompasses, or are the difficulties associated with some situations?

Through preliminary interviews with second year university students specialising in science, two lines of reasoning were identified as constituting obstacles; relevant problematic situations were then proposed.

1. Students often reason as though a cause existed only when there is an obvious effect (chapter 4). This way of thinking leads many to refuse the possibility that a field may exist at a point where charges cannot move. Hence the first series of questionnaires presented below, which involve insulators: the aim is to call into question the idea that all charge contributes to the field, even in a context that is mistakenly thought to be unfavourable to the electric field.

2. The other aspect of reasoning studied here is the attribution of a causal status to "formulae": a quantity, X, mentioned in the algebraic expression of another quantity G=f(X), is interpreted as an exclusive "cause" of the phenomenon associated with that quantity G. Therefore, the contribution to the total electric field of charges whose value is not explicitly included in the expression of that field is not taken into account. Hence the second series of questionnaires asks about the total field **E** near a conductor, which is expressed by a formula which contains the surface charge density σ present on the conductor near the point being considered ($\mathbf{E}=(\sigma/\varepsilon_0)\,\mathbf{n}$, see box 2). It may seem strange that it should be possible to express a field whose sources are all the charges in the universe through such a local variable: in fact, that variable itself reflects the locations of all the charges present in the situation, since the distribution of charges on the conductor is determined by the position of all the other charges.

The students questioned (1300 in all) have all completed their secondary schooling and are in preparatory classes or at university in France, or at the Stockholm Polytechnic. They are grouped in four categories in the analysis of the results, according to the instruction they have received on the subject:

[1] The idea underlying this statement is not limited to the particular context of electric fields, it is highly important in physics.

from a scientific baccalaureate level (G1) to a complete course on electromagnetism (G4).[2]

Box 2

Field near a conductor and Gauss's theorem

The electric field at a point M located very near[a] a conductor with a surface charge density σ (at a point P, very near M) is given by the relationship **E**=σ/ε₀**n**, where **n** is the normal unit vector at P to the surface of the conducting material, directed outwards, and ε_0 is the permittivity of free space. This expression is valid in a vacuum, irrespective of the presence or absence of external charges.

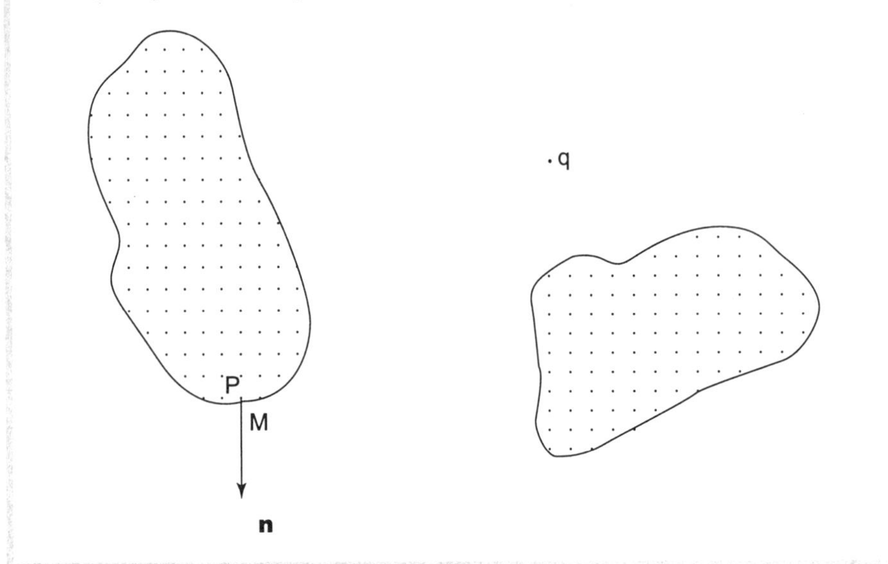

a. i.e., the solid angle subtended at M by the conductor of surface density σ is very close to 2π steradian; in other words, seen from M, it fills half the space.

[2] The students questioned had received instruction on the following notions, respectively:

(G1): electric circuits and an introduction to electrostatics; the electric field **E** defined by the relationship **F**=q**E**; the case of a uniform field, related to the capacitor, and the relationship E=U/d (Where U is the voltage and d the distance between plates);

(G2): (in addition to G1:) Coulomb's law and Gauss's theorem; the potential (V) defined by the relationship **E**=-**grad** V; the conductivity γ defined by the relationship **j**=y**E** (where **j** is the current density); a few rudiments of vector calculus, including the definition of a gradient;

(G3): (in addition to G2:) conductors in electrostatic equilibrium, the fact that the electric field is zero inside such conductors; field near a conductor;

(G4): (in addition to G3:) dielectrics (or "insulators") and Maxwell's equations.

2. "FIELD ONLY IF MOBILITY"? QUESTIONNAIRES ON INSULATORS

Two questionnaires were used for this study: the "insulator" questionnaire and the "Neapolitan ice cream" questionnaire.

2.1 The "insulator" questionnaire (INS)

The two items of this questionnaire raise the question of the action, inside an insulator, of an external charge, and of the action, outside the insulator, of an internal charge. In both cases, by virtue of the superposition principle, the correct answer is that the charge creates a non-zero electric field at the point considered.

INS 1:

A point charge is outside an insulating body. Does this charge create an electric field at a point M inside this insulating body (see diagram below)?

INS 2:

A point charge is inside an insulating body. Does this charge create an electric field at a point M outside this insulating body (see diagram below)?

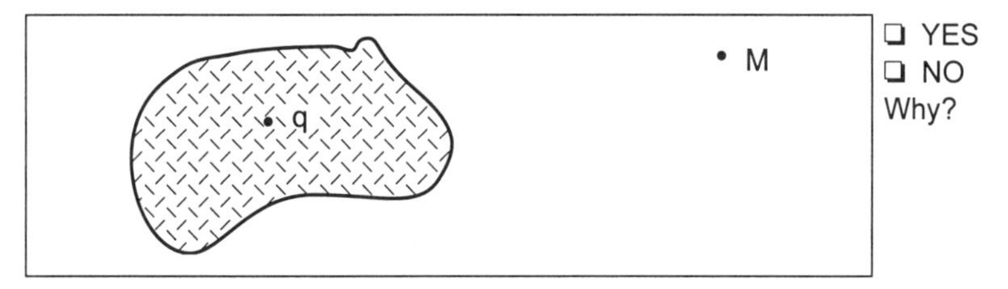

2.2 Main findings for the INS questionnaire

Table 1 presents the rate of answers ("YES", "NO", "I do not know," or "?" when no answer is provided). The category "E=0" was introduced because this indirect answer is not as categorical as a "NO": it might mean that the total field is zero, but that the contribution of the charge q is not negated for all that.

Table 1
Percentage of answers to the "insulator" questionnaire (INS)

Student group	N	INS 1				INS 2			
		YES	NO	E=0	?	YES	NO	E=0	?
G1	64	42	42	2	14	27	64	0	9
G2	25	36	48	0	16	60	20	4	16
G4	64	64	23	8	5	67	25	0	8

The rate of wrong answers is relatively high for each group, considering the apparent simplicity of the question. It changes predictably with the level of education.

But the students' comments are good indicators of the reasoning that they apply.

In addition to the justifications of correct answers ("YES, the charge creates a field"), the comments fall into two main categories:

The first category consists of answers alluding to the "blocking role of the insulator":

> "The insulating property of the body prevents the field from penetrating." (INS 1, G1)

> "The insulator is globally subjected to the electric field created by q. But M is not subjected to the electric field because it is protected by the insulator." (INS 1, G1)

> "The insulator blocks the field inside the body." (INS 2, G1).

The second category of comments all specify that the charges cannot move:

> "As this body is an insulator, the charges inside it are immovable. If q created a field, the charges inside would be subjected to an electric field." (INS 1, G2)

"As there is no free charge around M, there can be no electric field at M." (INS 1, G4)

Sometimes, the mobility required for a field to exist is curiously associated to the circulation of mediating particles:

"Protons or electrons emitted by q cannot reach M inside the insulator because that body insulates it." (INS 1, G1).

Equally surprising are answers which associate the condition of mobility with the charge that is thought to create the field:

"As it is not possible for charges to move inside the insulator, there can be no field created at M (outside)." (INS 2, G4)

As it is difficult to distinguish between the many different answers of this type, they have been grouped under a single heading: "field only if mobility."

Table 2 gives an idea of the frequency with which these two types of comments appear for one or the other of the two questions, or for both.

These two categories of comments are mutually exclusive as regards the process of classifying the justifications, since the non-mobility of charges is explicitly mentioned only in the first case . But the very fact of referring to the blocking property of the insulator is not far from suggesting this idea. Grouping all the comments which focus in a more or less explicit fashion on the idea that an insulator makes electrostatic interaction between charges impossible shows that a considerable proportion of students think along these lines: between 69% and 33%, depending on the level of education.

Table 2

Two types of comments for the INS questionnaire, and percentage of answers.

Students	Blocking insulator			Field only if mobility		
	INS 1	INS 2	INS 1 or INS 2	INS 1	INS 2	INS 1 or INS 2
G1 (N=64)	34	37	44	16	14	25
G2 (N=25)	8	20	28	16	4	20
G4 (N=64)	9	9	18	8	5	13

A second questionnaire deals with the same difficulty.

2.3 The "Neapolitan ice cream" questionnaire (NIC)

This questionnaire was devised to confront students with the idea that identical electric fields can have very different effects depending on the mobility of charges:

In the situation represented in the drawing below, the two equipotential surfaces are infinite planes, perpendicular to the plane of the figure. Describe the electric field between the two surfaces.

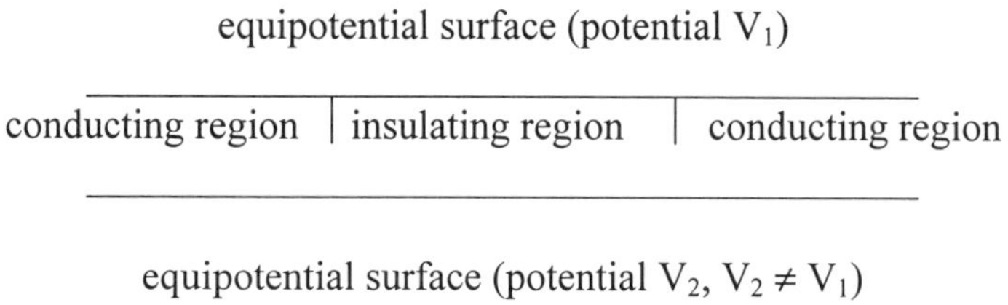

equipotential surface (potential V_1)

conducting region | insulating region | conducting region

equipotential surface (potential V_2, $V_2 \neq V_1$)

In your view, are the electric fields in the conducting region and in the insulating region…

different?

equal?

I do not know

Justify your answer.

The correct answer is that the fields are equal (if we neglect the edge effects).[3]

[3] The only correct argument that the students in G1 have at their disposal is the relationship E=U/d (where the voltage U and distance d between the plates are the same for the two materials). At more advanced levels, they can argue that the equipotential surfaces are the same, and that the relationship **E**=-**grad** V applies to any material. Finally, the most advanced students can base their reasoning on the continuity of the tangential component of the electric field **E** at the boundary between the insulator and the conductor.

2.4 Main findings for the "NIC" questionnaire

The principal answers are given in table 3.

Table 3
Percentages of answers for the "NIC" questionnaire

Students	Number	$E_{ins}=E_{cond}$ (correct)	$E_{ins}\neq E_{cond}$	I do not know
G1	185	26%	57%	17%
G2	80	49%	44%	7%
G3	77	27%	60%	13%
G4	78	18%	68%	14%

Nearly all the correct answers are based on the relationship E=U/d and on the fact that U and d are equal in the two zones.

As for the justifications of incorrect answers, $E_{ins}\neq E_{cond}$, they fall into four categories:

- $E_{cond}=0$: the students say that "the field is zero inside a conductor", without always specifying, that this is only true at electrostatic equilibrium:

  "For the conductor, the field inside is inevitably zero. The field that is created inside is zero because the potential is constant." (G3)

- Insulator: these justifications are based on the idea of an insulator, with or without a specific reference to the mobility of charges:

  "The electric field is only present in the conducting regions. The role of the insulator is to insulate, as its name suggests, and therefore it dampens or suppresses the field." (G2)

  "Since an insulator does not conduct electricity, the field cannot pass." (G1)

- Current: current is alluded to as passing through a conductor, but not through an insulator:

  "In the insulator, j=0, in the conductor, $j=\rho v=\gamma E$, and so the electric fields are different," or "Since the electric field is responsible for the current and since the currents in the conductor and in the insulator are different, the fields are different." (G2)

- Material: the students refer to "relative permittivity" or to "different conductivities," or simply to differences in the materials.

Some answers, like the "current" answer above, fall into several of these categories. For this reason table 4 includes a column grouping "insulator," "current" and "material" justifications (I/C/M).

Table 4
Types and rates of justifications for the answer "$\mathbf{E}_{ins} \neq \mathbf{E}_{con}$"

Students	Number	$\mathbf{E}_{ins} \neq \mathbf{E}_{cond}$	$\mathbf{E}_{cond}=\mathbf{0}$	Insulator	Current	Material	I/C/M	No answer
G1	185	57%	3%	25%	13%	10%	43%	12%
G2	80	44%	1%	16%	4%	15%	34%	9%
G3	77	60%	18%	17%	5%	16%	31%	13%
G4	78	68%	19%	3%	9%	29%	41%	8%

The high rate of incorrect answers ("NO, the field is not equal in the two zones") confirms that it is difficult to envisage the electric field independently of its effects, which are linked to the greater or lesser mobility of the charges. The comments grouped under the heading I/C/M in table 4 indicate that a major concern is, "How does it pass?" Some students mean the current, others mean the field. And for other groups, "it" designates an ill-defined quantity, "electricity," perhaps. Sometimes, an unspecified blockage becomes apparent:

> "I don't feel right saying that it is the same field, because there ought to be a difference between insulators and conductors."

The most frequent argument remains: "No current, therefore no field".[4]

Here one finds confirmation of the first type of obstacle linked to the application of the superposition principle: without mobility of charges, without an obvious effect of the electric field, it is difficult to admit the existence of the field. No effect, no cause – that was apparently the watchword of the teenagers who, when questioned, denied that air acts on one side of a piston, in the absence of motion (Séré, 1985); enduring trends

[4] Some remarks on the influence of the level of education:
It is more common for (G3) and (G4) students to justify their answers by saying that "The field is zero inside a conductor" (about 20%, as opposed to 3% for (G1) students). This is probably attributable to the fact that these students study conductors in equilibrium and "forget" about nonequilibrium situations. Another difference is the greater number of justifications in terms of "materials" among (G4) students (29% as opposed to 16% elsewhere). The (G4) students have recently studied dielectrics and the notion of permittivity, and no longer consider insulators as obstacles to the field (3% as opposed to at least 16% elsewhere). Thus, each level of education leaves its mark, but the result is a sort of conceptual layer cake: at the end of their studies, students are far from having a unified vision of the electric field.

of common sense persist at high levels of instruction, even on subjects that are highly theorised.

3. CAUSE IN THE FORMULA: THE "σ/ε_0" QUESTIONNAIRE

Let us move on to the second difficulty that students encounter, which is linked to an improper causal interpretation of formulae. This study deals with the expression for an electric field in the vicinity of a conductor, as defined in Coulomb's theorem (box 2). It seeks to determine whether the formula $\mathbf{E}=\sigma/\varepsilon_0\mathbf{n}$ makes it difficult to grasp the idea that the sources of the field are not all located on the conductor, even though the only charges mentioned in the formula are located on it. A correct understanding of the situation means recognising that a charge outside the conductor has two effects. The first is that the surface charge density σ on the conductor is modified by its influence. The second is a direct effect – the contribution to the total field at every point by virtue of the superposition principle. What follows is a questionnaire bearing on that point.

3.1 The σ/ε_0 questionnaire

At a point P on the surface of a conductor at equilibrium, the surface charge density is σ. $\mathbf{n}$ is defined as the normal unit vector perpendicular to the surface at P, pointing outwards. Outside the conductor, at a point very near P, the electric field is $\mathbf{E}=(\sigma/\varepsilon_0)\mathbf{n}$.

1. Is this field due…

a) to the charges in the vicinity of P?

b) to all the charges on the conductor?

c) to all the charges in the universe?

Choose the right answer and justify.

2. Consider the two situations below:

$\mathbf{n}$ is defined as the unit vector perpendicular to the conductor at P, pointing outwards.

a) In situation B, will the electric field $\mathbf{E}$ outside the conductor at a point very near P be given by the formula $\mathbf{E}=(\sigma/\varepsilon_0)\mathbf{n}$?

Yes No I do not know

Justify your answer.

b) Is the electric field $\mathbf{E}$ outside the conductor, at a point very near P, the same in situations A and B?

Yes No I do not know

Justify your answer.

c) Is the surface charge density σ the same in situations A and B?
Yes No I do not know
Justify your answer.

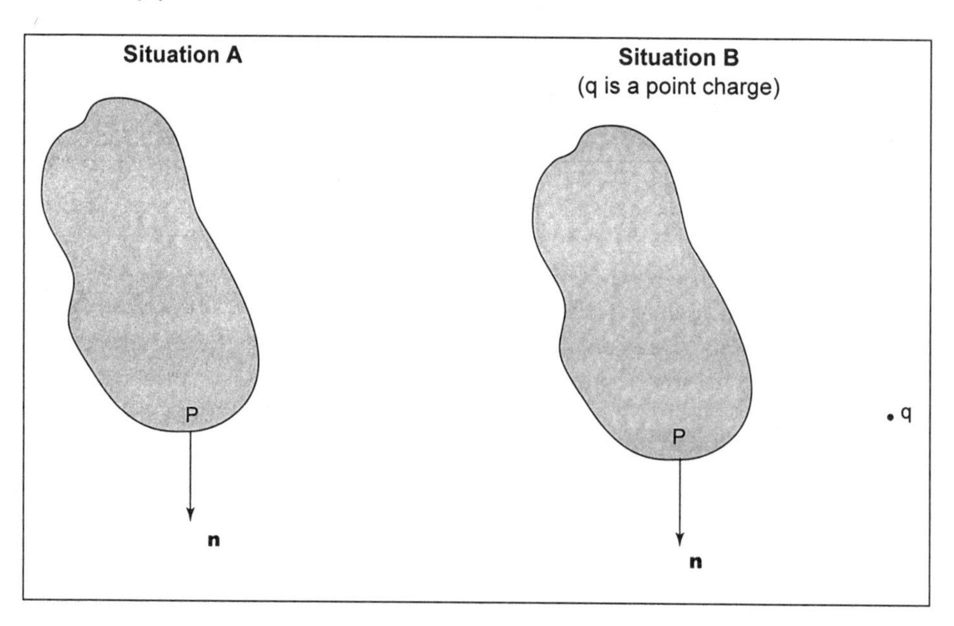

Correct answers:

The formula $\mathbf{E}=(\sigma/\varepsilon_0)\mathbf{n}$ is an expression for the total field, created by all the charges in the universe. It is valid no matter what the environment of the conductor (answer: YES for question 2a). And the values of the quantities E and σ_p are affected by the presence of the external charge (answers: NO for questions 2b and 2c).

3.2 Main findings for the"σ/ε_0" questionnaire

Tables 5 to 8 give the students' answers.

Table 5
Question 1 : $\sigma/\varepsilon_0\mathbf{n}$ is created by…

Students	Number	…the charges in the universe (correct)	...the conductor	…the vicinity	I do not know
G3	194	30%	44%	24%	2%
G4	58	9%	60%	29%	2%

Table 6
Question 2a: In situation B, does the formula $\mathbf{E}=(\sigma/\varepsilon_0)\mathbf{n}$ always hold?

Students	Number	YES (correct)	NO	I do not know
G3	194	51%	44%	5%
G4	58	29%	57%	14%

Table 7
Question 2b: Is the field the same in situations A and B?

Students	Number	NO (correct)	YES	I do not know
G3	194	78%	18%	4%
G4	58	78%	15%	7%

Table 8
Question 2c: Is σ the same in situations A and B?

Students	Number	NO (correct)	YES	I do not know
G3	194	65%	32%	3%
G4	58	52%	40%	8%

It is noticeable that, with the exception of question 2b, concerning the field **E**, these items elicit many erroneous answers. The most frequent errors concern the sources of the field, often limited to the conductor, and the validity of the formula, which is questioned or denied by 49% and 71% of the population, respectively (G3, G4). Here is an example of a formula that is well known to the students, but whose true meaning escapes at least half of them.

Few comments are provided with the answers on the sources of the field (table 5). Some students mention the distance of the external charges as a justification for their non-intervention, but that may be a way to avoid committing themselves with a more precise statement. More often, the formula itself is questioned, and considered as relating solely to the field produced by the conductor:

"All the charges on the conductor will contribute. All the charges in the universe do not enter into the formula $\sigma/\varepsilon_0\mathbf{n}$." (G4)

"All the charges ought to contribute to the field outside the conductor. Since the charge density σ appears in the formula $\sigma/\varepsilon_0\mathbf{n}$, the origin of the field is in the conductor." (G4)

"In the vicinity of P, the surface charge density is σ; it will create $\sigma/\varepsilon_0\mathbf{n}$." (G3)

In justifications concerning the second part of the questionnaire (table 9), too, we again find the idea that the effect of the external charge is negligible. We also find the idea of influence (an indirect effect of q), although the direct contribution is not always made explicit:

"The formula is true... Because of the influence of q, $\sigma_A \neq \sigma_B$, so, because of the formula, $\mathbf{E}_A \neq \mathbf{E}_B$." (G4)

Table 9
"σ/ε_0" questionnaire: justifications for answers to question 2 and rates of response.

Students	Number	None	Effect of an external charge: negligible	Influence only (direct effect is not taken into account)	Super-position only (influence is not taken into account)	Influence + super-position	Other
G3	194	10%	2%	31%	29%	18%	10%
G4	58	9%	7%	22%	38%	17%	7%

Other comments mention that the contributions of the charges of the conductor and of the external charge q are superposed; the relationships provided are often erroneous (whether or not influence is considered):

"q creates another field $\mathbf{E'}$. In situation A, $\mathbf{E}=\sigma/\varepsilon_0\mathbf{n}$ is created by the environment. In situation B, $\mathbf{E}_T=\mathbf{E}+\mathbf{E'}$." (G3)

"$\mathbf{E}=\sigma/\varepsilon_0\mathbf{n}+\mathbf{n}_q/4\pi\varepsilon_0 r^2$." (G3)

"$\mathbf{E}=\sigma/\varepsilon_0\mathbf{n}+\mathbf{E}q$." (G3)

In these answers, it seems that the first term is the supposed contribution of the conductor and the second term that of the external charge. The correct formula, $\mathbf{E}=\sigma/\varepsilon_0\mathbf{n}$, has, therefore, been reinterpreted, and the field (on the left hand side) has been attributed to the charges of the conductor (on the right hand side). Here, the difficulty is not that the principle of superposition is neglected, but that the students do not know that it is already incorporated into the formula, so that this becomes misleading if interpreted in terms of cause and effect. The results of a test question shown in box 3 confirm the extent of this difficulty.

Box 3

The difficulty of taking into account a superposition that is "invisible" in the formula

A test question asks students to assess the following (erroneous) relationship for the situation shown in the diagram, which has already been described.

A student suggests the expression $\mathbf{E}(M)=\sigma/\varepsilon_0\mathbf{n}+\mathbf{E}_q$, where $\mathbf{E}_q$ is the field created by the charge q at a point M very near a conductor.

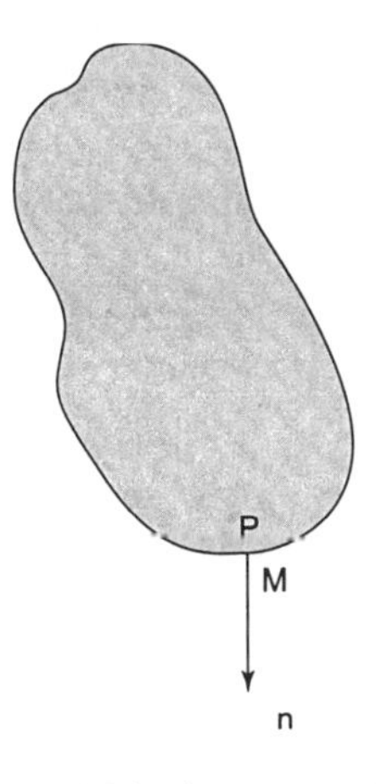

What do you think? Is the expression...

right? YES NO Explain why

wrong? YES NO Explain why and give the correct expression

ambiguous? YES NO Explain why and give an unambiguous formula

One third of the students at levels G3 (N=74) and G4 (N=67) state that the relationship is correct, and three quarters show by their justifications to right or wrong answers that they do not understand the situation, as can be seen from the following commentary, provided with a correct answer ("the formula is incorrect"): "As σ changes, one ought to write $\mathbf{E}(M)=\sigma'/\varepsilon_0\mathbf{n}+\mathbf{E}_q$."

4. SUMMARY AND PEDAGOGICAL PERSPECTIVES

We can see that the two types of obstacle identified in the preliminary inquiry do in fact hinder understanding. Both can be traced back to an improper, causal analysis.

For many students, cause and effect are associated in an overly restricting relationship. This leads them to deny that there is a component of the electrical field in insulators, when there is no obvious effect. More generally, we might wonder whether, for these students, the electric field in electrostatics has anything to do with the electric field in electrodynamics (see also Benseghir and Closset, 1993, and chapter 11).

As regards the restricted interpretation of the expression for a quantity in terms of cause and effect, the use of Gauss's theorem for the field near a conductor can shed light on this tendency, provided, as always, that the right questions are asked. Unless one tries it, one might think that the formulation of that generally well-known theorem is perfectly clear. But the students' acceptance of an erroneous formula shows how little they really understand. And this is true even for one of the most reputable Mathématiques Spéciales classes in France.

Are these obstacles then insuperable? That could only be proved if they were explicitly targeted in teaching, which is not the case in France today.

Some noteworthy results have already been obtained through experiments in a class of Mathématiques Spéciales Technologiques (Rainson, 1995). Two relatively unsuccessful attempts to overcome these obstacles (1991-1992 and 1992-1993) confirmed the fact that a few assertions and adequate examples will not suffice. The teaching strategies that finally had notable effects (1993-1994, though there was no miraculous progress: a 20% to 40% rise in correct answer rates for diagnostic questions; these results were confirmed and stable over the three next years – 1994, 1995, 1996) were those suggested by the above analysis:

- Explicit work on causes – the sources of the field – was undertaken, using a visualisation technique based on superposed transparencies: one transparency for the charge, another for the conductor that is influenced. This technique makes it possible to illustrate the direct role of a charge, whose permanent association with its field is materialised by a given transparency. The indirect role of the charge, via influence, is also visualised for each position of that charge relative to the conductor, by a representation of the surface charge density on the conductor.
- This illustrative approach was adopted for situations in which influence occurs and causes changes which are analysed step by step. Indeed, transformation situations tend to encourage the causal analysis of

phenomena (see chapter 5, notably). The aim is to solicit a natural approach in terms of cause and effect, in order to evaluate explicitly its possibilities and limitations.

Here, as elsewhere, the analysis of common difficulties does not give rise to only depressing assessments. It can also orientate decisions on objectives and suggest what action should be taken (Viennot and Rainson, 1999).

REFERENCES

Benseghir, A. and Closset, J.L. 1993. Transition électrostatique électrocinétique: points de vue historique et didactique, *Didaskalia* 2, pp 31-47.

Rainson, S. 1995. *Superposition des champs électriques et causalité: étude de raisonnements, élaboration et évaluation d'une intervention pédagogique en classe de Mathématiques Spéciales Technologiques* Thesis, Université Paris 7.

Rainson, S., Tranströmer, G. and Viennot, L. 1994. Students' understanding of superposition of electric fields. *American Journal of Physics*, 62 (11), pp 1026-1032.

Séré, M.G. 1985. *Analyse des conceptions de l'état gazeux qu'ont les enfants de 11 à 13 ans, en liaison avec la notion de pression, et propositions de stratégies pédagogiques pour en faciliter l'évolution.* Thesis (Thèse de doctorat d'état). Université Paris 6.

Viennot, L. and Rainson, S. 1992. Students' reasoning about the superposition of electric fields, *International Journal of Science Education.* 14 (4), pp 475-487.

Viennot, L. and Rainson, S. 1999. Design and evaluation of a research-based teaching sequence: The superposition of electric fields, *International Journal of Science Education, Special issue: Conceptual Development in Science Education*, vol 21 (1), pp 1-16.

Conclusion

This short book provides many results on common trends of reasoning. Its aim was to show their consistency. And indeed, the search for fundamental threads that orientate students' responses has been fruitful.

The central idea that orientates common reasoning is that of a quasi-material object, endowed with intrinsic properties. Thus, the colour of an object is naturally understood as independent of the light which illuminates it. Similarly, the velocity of an object is understood as independent of a frame of reference, and as being the effect of a motor. For several possible values to be admitted for this quantity, drag is thought to be necessary, as the actions of two motors then add. An optical image or a luminous ray is most often understood as a material object, the former travelling as a whole and the latter being visible from the side.

Common reasoning about the object being considered is always guided by a search for simplicity. Whereas accepted physics introduces different quantities – force, energy, and momentum – common reasoning uses a single notion, which can have various labels. When the considered object is "electricity," for instance, one finds that many arguments are based on a conglomerate of undifferentiated notions, i.e., current-voltage-energy.

Physical laws then take on a meaning that is remote from accepted theory. The fundamental relationship of dynamics, which relates force to acceleration, is commonly interpreted as establishing a direct link between motion – in fact, its velocity – and its cause. If this cause can be found only in the past, it is imagined as "stored" in the moving object, as a supply of force. Newton's second law is then readily used as a direct link between former force and present speed, and the fact that the quantities involved in this relationship are to be taken "at the same instant" is ignored. The same type of misapplication can be seen concerning the law of reciprocal actions (Newton's third law), which is often reinterpreted as expressing a dynamic conflict between objects, each of which is seen as having "its own force", resulting in a motion of the two objects in the direction of the largest force – the law is then violated. These interpretations also involve temporal shifts: the simultaneity of action and reaction does not appear as inevitable.

Simultaneity is also a stumbling block as regards the proper understanding of the motion of complex systems. When many quantities are changing simultaneously, it is very often possible to carry out a "quasi-static" or quasistationary analysis, where some simple laws are considered as permanently valid. To say this is equivalent to saying that the different parts of the systems, and/or the different quantities, "warn" each other of the perturbations that occur, instantaneously (on the time-scale being considered). On the other hand, common reasoning uses a linear sequence of cause/effect relationships, each of these terms being associated with a single quantity. "Then... then..." is a classical pattern of arguments, time shifts being surreptitiously introduced between the changes in the quantities involved. Simultaneous changes constrained by permanent laws are explained in terms of a story. In this type of reasoning, not only are variables considered "one at a time," but a series of episodes is imagined. This chronology justifies all kinds of assertions that would be patently contradictory if simultaneity were really envisaged. We can probably safely assume that such a structure of reasoning is very general. Who has not recognised some of its features in the arguments commonly put forward in economics, in environmental studies, or in biology?

Objects and stories: if one had to sum up common reasoning in three words, that description would suffice. It is necessary, however, to go into more detail to show that one cannot manage in physics if one remains within that register, and that there is, in fact, a definite coherence in the responses this reasoning leads to.

Of course, error should not always be seen as momentous, in one of those swings of the pendulum that too often characterises thinking about education. Every wrong answer is not the tip of an iceberg of common thought. Inattentiveness, random mistakes in calculations, answers shaped to suit the teacher's expectations, and "Who knows what else", may also be behind learners' mistakes. Moreover, the analysis proposed here does not always determine precisely how much is "meaningful" in a given individual response, and how much of it is "Who knows what." But anyone who realises the coherence of common thought will have a very different view of common arguments. Features that had gone unnoticed suddenly stand out. Insignificant answers become profound. The explanations that one referred to most readily when teaching are seen as calling for attention and re-evaluation.

Would it be better not to know all this, and to go on teaching regardless, complaining that students do not understand anything because they have not worked hard enough in the preceding grades, and are not working any harder now? Certainly not.

How can one deny that having a key with which to read students' productions leads to pedagogical interaction that is more focused on crucial points, and therefore more efficient?

This applies to higher levels of education as well. In its annual report, the jury of a competitive examination for teachers (France, CAPES 1994) cites one far-too-frequent mistake. On the subject of acoustic propagation through the interface of two different media, the candidates frequently write that the excess pressure of the incident wave is equal to the sum of the excess pressures of the transmitted and reflected waves (or they make the same mistake for the flow rate). "A very serious mistake," says the report. But is reproof sufficient? Should we not try to analyse the difficulty in order to intervene in a positive manner? Yes, it is a mistake: in accepted theory, the quantities relating to each side, respectively, have to be added up, then equated (excess pressures of the incident and reflected waves on one side, excess pressure of the transmitted wave on the other). The correct analysis is determined by space, whereas it is chronology that inspires the erroneous response. Common reasoning follows the progress and division at the interface of a wave, and envisages the quantity "excess pressure," in a rather substantial form. An analysis of this kind, in which accepted theory and a common mistake are set alongside one another, may help students to understand the situation better.

Moreover, an analysis of natural reasoning enables us to be more vigilant as regards the texts or documents that we offer to our students, be they written by ourselves or by others. This may also lead our students to be more vigilant about the innumerable popularisations they come across – books or magazines that appeal to natural reasoning and constantly reinforce those aspects of it which are most remote from accepted theory.

However, the conclusion of this text should not be read as a call to rally against error in physics. In fact, the description proposed here, which traces some of the main lines of natural reasoning, offers a positive perspective on the fight against error. Errors become interesting, because they reveal a thought process and, by contrast, shed light on some structural features of accepted theory. The stumbling blocks in common reasoning point to certain aspects of theory on which particular stress should be laid.

This leads to the positive side of such an analysis, which emphasises how non-obvious, and therefore how interesting, some aspects of physics can be, despite the fact that they are now termed "elementary," and in spite of familiarity and ambiguous formulations.

Thus, greater awareness and more adequate goal setting are complementary.

But these can be applied in a more or less radical manner.

Indeed, once the difficulties and goals have been identified for each topic, it seems natural to use questionnaires such as the ones presented in this book as tools for teaching. Anyone who has participated in these highly motivating questioning situations, and in the heated discussions that follow among different and often somewhat blasé student populations, will understand why, afterwards, one would like to apply this technique everywhere.

We also need to envisage local strategies and pedagogical tools adapted to each particular case.

However, this may put us at some distance from classical teaching, i.e., at odds with the views that prevailed in the eighties, and that form the context of the inquiries reported in this book: "Considering all they've been taught, they ought at least to realise that...." Whether it is seen as a luxury, as a sign of morbid curiosity, or as a helpful indicator, concern as to whether "elementary" notions have in fact been understood appears slightly exotic. And all the more so as many teachers believe that if "elementary" notions are not properly understood, this is because students have not been taught how to manipulate formulae properly.

A much more radical conclusion would be that the "elementary" aspects of physics should be introduced, at least along some of the lines identified here, at an early stage in education, and as explicitly and simply as possible.

Of course, the theory itself is explicit, and, most of the time, teachers do present physical hypotheses, the mechanisms used to elaborate the theory, and the conclusions that can be drawn from these, in an unambiguous fashion. And yet, it is necessary to be more explicit still, in order to underline some aspects that are implicit in the theory but that only experts think of seeing.

For instance, when a variable is missing from a relationship, this means an independence which, in some cases, is anything but trivial. For instance, the speed of sound (c) in a gas (of given molar mass M and given coefficient γ) at a given absolute temperature T, does not depend on its loudness: you can shout louder, but the sound will not travel faster. Not to stress this point means not letting the relationship $c=(\gamma RT/M)^{1/2}$ say all that it has to say.

This remark applies to many common statements which include the word "constant," and are loaded with implicit meaning. As regards the speed of light in a vacuum, for instance, students are not generally aware of the most astonishing independence, namely the frame of reference. The speed is not affected by the frequency of the radiation, either – another point which is far from obvious. The interesting independences, as a result of which we give some quantities the label of "constant", therefore deserve to be formulated explicitly.

More generally, relationships between quantities are not only a means of calculating certain numerical values from given values others. They express functional dependences which can be at least partly expressed in words. So, in addition to the independence commented on above, the expression for the speed of sound makes it possible to say that this speed decreases, for a given gas, when the temperature decreases: the sound barrier is crossed at a lower speed at higher altitudes because the air is colder there, not because the density of the particles is lower.

Developing this ability to read relationships functionally is absolutely essential for evaluating the validity of results, estimating their significance, confronting their consequences with those suggested by intuition; in brief, it gives meaning to physics.

But it is not necessary to go into everything all at once. Relevant aspects can be selected even within a given relationship. To come back to the example of Newton's second law, **F**=m**a**, it can be applied even in the following, reduced form: Depending on whether or not the velocity is changed with time, the total force exerted on the moving object is, respectively, non-zero or zero. Just understanding this fact requires a great deal of work. The whole content of the relationship is not there, but this makes it possible to do some work without going too far into kinematics, and using lots of "$1/2gt^2$".

This approach was chosen for grade 11, in France, in 1993.[1] Not that the mastery of calculus is considered unimportant – it is necessary for future scientists. The point is to identify the essential meaning of a relationship, independently of technical obstacles, in the knowledge that the finer points can be addressed later on. In other words, it is the exact opposite of what is commonly thought necessary, that is, that one should do a great deal of calculus before one can hope to understand anything. Everyone will agree that it is better to be able to do it all, detailed calculations as well as verbal syntheses. But teaching demands a degree of realism, it has to be adapted to a given population, difficult choices must be made, a progression charted, and goals must be negotiated. At the first grade levels in which physics is taught, presenting reduced laws may be a valuable means of introducing as much meaning as possible at the least cost; and it reduces the risk of leaving the students totally at sea.

Anyone fearing that this will make physics too easy has no cause for alarm. Accepting that it is necessary to take several causes, variables and

[1] Première S, i.e., the eleventh year of secondary education (science section).

systems into account when analysing a phenomenon, correctly applying a law (even a reduced law), situating the quantities it involves in time, interpreting results, confronting possible contradictions, determining how far an argument will go, knowing when to conclude that it is not possible to conclude: these are some of the many scientific abilities linked with the rigorous practice of qualitative reasoning.

That is why this book ends with a plea for the teaching of reasoning in physics.

It cannot be pointed out too often that it takes more than experiments to make students really active, that seeing does not mean understanding, that calculating does not necessarily entail a mastery of formalism. In all these activities, greater conceptual knowledge crucially depends on explicitness and linkage. Learners approaching a text out of curiosity or interest, or simply because they have to, must know how to find meaning and coherence there. Otherwise, there is no pleasure in learning, and nothing much can happen. But this requires effort.

The efforts that we make towards developing our students' determination to understand and their demand for coherence will, we hope, lead them to seek meaning and rigour in other fields as well. However, the context in which such efforts are undertaken is unavoidably complex; other respectable objectives and numerous constraints have to be taken into account. Negotiation is, therefore, inevitable, but there are more choices open to us than is generally allowed.

In any case, beliefs concerning teaching must not go unquestioned. The days of "all-it-takes-is" are over. What we now need, in order to determine the pedagogical uses to which the results summed up in this book can be put, are pertinent assessments of the various efforts made at local and national level. Defining teaching is a multi-variable problem if there ever was one. Adjusting each variable to all the others is an intricate process, and it cannot be accomplished through dogmatism of any kind. What is to be desired, however, is that every decision-maker, including the individual teacher or student, should have at his or her disposal as much information as possible. I hope that this book will prove useful in that respect.

Bibliography

Andersson, B. 1986. The experiential Gestalt of Causation: a common core to pupils preconceptions in science. *European Journal of Science Education*, 8 (2), pp 155-171.

Artigue, M., Ménigaux, J. and Viennot, L. 1990. Some aspects of students' conceptions and difficulties about differentials. *European Journal of Physics*, 11, pp 262-267.

Artigue, M., Ménigaux, J. and Viennot, L. 1991. *Questionnaires de travail sur les différentielles*. Université Paris 7 (Diffusion IREM et LDPES).

Bachelard, G. 1938. *La formation de l'esprit scientifique*, Vrin, Paris.

Bachelard, G. 1966. *Le rationalisme appliqué*, PUF Paris (1949).

Bauer, E. 1948. *L'électromagnétisme, hier et aujourd'hui.* Paris, Albin Michel.

Benseghir, A. 1989. *Transition électrostatique-électrocinétique : point de vue historique et analyse des difficultés des élèves.* Paris, Thesis, Université Paris 7.

Benseghir, A. and Closset, J.L. 1993. Prégnance de l'explication électrostatique dans la construction du concept de circuit électrique: points de vue historique et didactique. *Didaskalia* 2, pp 31-47.

Benseghir, A. and Closset, J.L. 1996. The electrostatic-electrokinetic transition: historical and educational difficulties, *International Journal of Science Education*, Vol 18, (2), pp 179-191.

Biot, J.B. 1801. Rapport fait à la classe des sciences mathématiques et physiques de l'institut national sur les expériences du cit. Volta. *A.C.*, n° 41, pp 3-23.

Blondel, C. 1982. *Ampère et la création de l'électrodynamique.* Paris, Bibliothèque Nationale.

Blondin, C., Closset, J.L. and Lafontaine, D. 1992. Raisonnements en électricité et en électrodynamique. *Aster* 14, pp 143-155.

Bovet, M., Greco, P., Papert, S. and Voyat, G. 1967. Perception et notion du temps, *Etudes d'épistémologie génétique*, vol XXI, Paris, P.U.F.

Brown, T. 1969. The electric current in early nineteenth century French physics. In *Historical Studies in the Physical Sciences*. vol. I, pp 61-103.

Bruno, G. 1584. La Cena de le Ceneri III, 5 *Opere Italiane*, (Ed; Wagner,1830).

Bulletin Officiel du Ministère de l'Education Nationale 1992a, n°31, *Programmes des classes de quatrième et quatrième technologique*, pp 2086-2112.

Bulletin Officiel du Ministère de l'Education Nationale, 1992b. *Nouveaux programmes de physique et chimie pour les classes de Seconde, Première, et Terminale des lycées*, Numéro hors série du 24-9-1992, Vol II, p 38.

Bulletin Officiel du Ministère de l'Education Nationale, 1993, n°93, *Nouveaux programmes de physique et chimie pour la classe de Troisième des collèges*, pp 3721-3737.

Caldas, E., 1994. *Le frottement solide sec: le frottement de glissement et de non glissement. Etude des difficultés des étudiants et analyse de manuels.*. Thesis. Université Paris 7.

Caldas, E. and Saltiel, E. 1995. Le frottement cinétique: analyse des raisonnements des étudiants. *Didaskalia,* 6, pp 55-71.

Chauvet, F. 1990. *Lumière et vision vues par des étudiants d'arts appliqués*, unpublished Mémoire de Tutorat (L.D.P.E.S.), D.E.A. in Didactics, Université Paris 7.

Chauvet, F. 1993. Conception et premiers essais d'une séquence sur la couleur, *Bulletin de l'Union des Physiciens* , 750, pp 1-28.

Chauvet, F. 1994. *Construction d'une compréhension de la couleur intégrant sciences, techniques et perception: principes d'élaboration et évaluation d'une séquence d'enseignement.* Thesis. Université Paris 7.

Chauvet, F. 1996. Teaching colour : designing and evaluation of a sequence, *European Journal of Teacher Education*, vol 19, n°2, pp 119-134.

Closset, J.-L. 1983. *Le raisonnement séquentiel en électrocinétique.* Paris, Thèse, Université Paris 7.

Closset, J.-L. 1989. Les obstacles à l'apprentissage de l'électrocinétique. *Bulletin de l'Union des Physiciens*, 716, pp 931-950.

Closset, J.L. and Viennot, L. 1984. Contribution du raisonnement naturel en physique. In Schiele, B. and Belisle, C. (Eds.): *Les représentations. Communication -Information* 6 (2-3), pp 399-420.

Couchouron, M., Viennot, L. and Courdille, J.M. 1996. Les habitudes des enseignants et les intentions didactiques des nouveaux programmes d'électricité de Quatrième, *Didaskalia* n°8, pp 83-99.

Courdille, J.M. 1991. *Questionnaires de travail sur l'électrocinétique* Université Paris 7 (IREM and LDPES).

Crepault, J. 1981. Etude longitudinale des inférences cinématiques chez le préadolescent et l'adolescent: évolution et régression, *Canadian Journal of Psychology*. 35,3.

Cuvier, G. 1801. Rapport sur le galvanisme. *Journal de Physique*, n° 52, pp 318-324.

De La Rive, A. 1825. Mémoire sur quelques uns des phénomènes que présente l'électricité voltaïque dans son passage à travers les conducteurs liquides. *A.C.P.*, n° 28, pp 190-221.

Di Sessa, A. 1988: Knowledge in pieces. In Formann, G. and Pufall, P. (Eds.) *Constructivism in the Computer Age*. Lawrence Erlbaum Associates, Hillside, NJ., pp 49-70.

Driver, R., Guesne, E. and Tiberghien, A. 1985. Some features of Children's Ideas and their Implications for Teaching. In Driver, R., Guesne, E. and Tiberghien, A. (eds): *Children's Ideas in Science*. Open University Press, Milton Keynes, pp 193-201.

Fauconnet, S. 1981. *Etude de résolution de problèmes: quelques problèmes de même structure en physique.* Thesis (Thèse de troisième cycle), Université Paris 7.

Fawaz, A. 1985. *Image optique et vision: étude exploratoire sur les difficultés des élèves de première au Liban.* Thesis (Thèse de troisième cycle), Université Paris 7.

Fawaz, A. and Viennot L. 1986. Image optique et vision, *Bulletin de l'Union des Physiciens* , 686, pp 1125-1146.

Feher, E. and Rice, K. 1987. A comparison of teacher-students conceptions in optics, *Proceedings of the Second International Seminar: Misconceptions and Educational Strategies in Science and Mathematics*, Cornell University, Vol II, pp 108-117.

Galili, Y. 1996. Students' conceptual change in geometrical optics, *International Journal of Science Education,* 18 (7), pp 847-868.

Galili, Y. and Hazan, A. 2000. Learners' knowledge in optics, *International Journal of Science Education,* 22 (1), pp 57-88.

Goldberg, F.M. and Mc Dermott, L. 1987. An investigation of students' understanding of the real image formed by a converging lens or concave mirror, *American Journal of Physics*, 55 (2), pp 108-119.

Greco, P. and Piaget, J. 1959. *Apprentissage et connaissance.* Paris, P.U.F.

Groupe Technique Disciplinaire de Physique 1992, *Document d'accompagnement pour la classe de Quatrième,* Ministère de l'Education Nationale et de la Culture.

Groupe Technique Disciplinaire de Physique 1992. *Document d'accompagnement du programme de Première*. Ministère de L'Education Nationale et de la Culture, Paris.

Groupe Technique Disciplinaire de Physique 1993. *Document d'accompagnement du programme de Troisième*. Ministère de L'Education Nationale et de la Culture, Paris.

Groupes Techniques Disciplinaires de Physique et de Chimie 1992. Avant-projets des programmes de physique et chimie, *Bulletin de l'Union des Physiciens* , 740, supplément pp 1-52.

Guesne, E. 1984. Children's ideas about light / les conceptions des enfants sur la lumière, *New Trends in Physics Teaching*, Vol IV UNESCO, Paris, pp 179-192.

Guesne, E., Tiberghien, A. and Delacôte, G. 1978. Méthodes et résultats concernant l'analyse des conceptions des élèves dans différents domaines de la physique. *Revue française de pédagogie*, 45, pp 25-32.

Gutierrez, R. and Ogborn, J. 1992. A causal framework for analysing alternative conceptions, *International Journal of Science Education.* 14 (2), pp 201-220.

Haüy, R.J. 1803. *Traité élémentaire de physique*, vol 1, 1st edition.

Hirn, C. 1995. Comment les enseignants de sciences physiques lisent-ils les intentions didactiques des nouveaux programmes d'optique de Quatrième? *Didaskalia*, 6, pp 39-54.

Hirn, C. and Viennot, L. 2000. Transformation of Didactic Intentions by Teachers: the Case of Geometrical Optics in Grade 8 in France, *International Journal of Science Education*, 22 (4), pp 357-384.

Inhelder, B. and Piaget, J. 1955. *De la logique de l'enfant à la logique de l'adolescent,* PUF, Paris.

Johsua,S. 1985. *Contribution à la délimitation du contraint et du possible dans l'enseignement de la physique (essai de didactique expérimentale),* Thesis (Thèse d'état), Marseille, Université de Provence.

Johsua, S. and Dupin, J.J. 1988. La gestion des contradictions dans les processus de modélisation en physique, en situation de classe. In *Didactique et acquisition des connaissances scientifiques, Actes du colloque de Sèvres, mai 1987*. Paris, Edition de la pensée sauvage.

Johsua, S. and Dupin, J.J. 1989. *Représentations et modélisations: le "débat scientifique" dans la classe et l'apprentissage de la physique*, Peter Lang, Berne.

Johsua, S. and Dupin, J.J. 1993. *Introduction à la didactique des sciences et des mathématiques,* P.U.F., Paris.

Jouguet, M. 1955. *Traité d'électricité théorique,* vol 2. Paris, Gauthiers-Villars.

Kaminski, W. 1989. Conceptions des enfants et des autres sur la lumière, *Bulletin de l'Union des Physiciens* , 716, pp 973-996.

Kaminski, W. 1991. *Optique élémentaire en classe de quatrième: raisons et impact sur les maîtres d'une maquette d'enseignement*, Thesis (L.D.P.E.S.), Université Paris 7.

Koyré, A. 1966. *Etudes Galiléennes* (p 136). Hermann Paris.

Kuhn, T.S. 1970 *The structure of scientific revolutions.* Chicago, IL: University of Chicago Press.

Lamé, G. 1837. *Cours de l'école polytechnique*, vol 2, 2ème partie.

Léna, P. and Blanchard, A. 1990. *Lumières. Une introduction aux phénomnes optiques.* Interéditions Paris.

Lijnse, P., Licht, P., de Vos, W. and Waarlo, A.J. 1990. *Relating Macroscopic Phenomena to Microscopic Particles, A Central Problem to secondary education.* CD-β Press Utrecht.

Maurines, L. 1986. *Premières notions sur la propagation des signaux mécaniques: étude des difficultés des étudiants*. Thesis. Université Paris 7.

Maurines, L. 1991. Raisonnement spontané sur la propagation des signaux: aspect fonctionnel. *Bulletin de l'Union des Physiciens*, 733, pp 669-677.

Maurines, L. 1993. Mécanique spontanée du son. *Trema*. IUFM de Montpellier, pp 77-91.

Maurines, L. and Saltiel, E. 1988a. Mécanique spontanée du signal. *Bulletin de l'Union des Physiciens*, 707, pp 1023-1041.

Maurines, L. and Saltiel, E. 1988b. *Questionnaires de travail sur la propagation d'un signal*, Université Paris 7 (diffusion LDPES)

Maury, L., Saltiel, E. and Viennot, L. 1977. Etude de la notion de mouvement chez l'enfant à partir des changements de repère, *Revue Française de Pédagogie*, 40, pp 15-29.

McDermott, L.C. 1984. Revue critique de la recherche dans le domaine de la mécanique.*Recherche en Didactique: les actes du premier atelier international, La Londe les Maures, 1993.* CNRS, Paris, pp137-182.

Méheut, M. 1994. Construction d'un modèle cinétique de gaz par les élèves de collège. Jeux de questionnement et de simulation. *Actes du Quatrième Séminaire National de la Didactique des Sciences Physiques*, IUFM d'Amiens, Amiens, pp 53-73.

Menigaux, J. 1986. Analyse des interactions en classe de troisième. *Bulletin de l'Union des Physiciens*, n° 683, pp 761-778.

Menigaux, J. 1991. Raisonnements des étudiants et des lycéens en mécanique du solide. *Bulletin de l'Union des Physiciens* n° 738, pp 1419-1429.

Michaud, Y. and Lemoal, Y. 1979. *Physique 4ème*. Paris, Magnard.

Moscovici, S. 1976. *La psychanalyse, son image et son public*. PUF, Paris.

Peltier, J.C.A. 1836. Courants électriques. Définition des expressions Quantité et Intensité. *Comptes rendus de l'académie des sciences,* n°2, pp 475-476.

Pfaff, C.H. 1829. Défense de la théorie de Volta relative à la production de l'électricité par le simple contact, contre les objections de M. le professeur De La Rive. *A.C.P.*, n° 41, pp. 236-247.

Piaget, J. 1975. *L'équilibration des structures cognitives, problème central de développement*. Etudes d'épistémologie génétique XXXIII, P.U.F., Paris.

Polvani, G. 1949. L'invention de la pile. *Revue de l'histoire des sciences*, n° 2, pp 340-351.

Posner, G., Strike K., Hewson, P. and Gertzog, W. 1982. Accomodation of a scientific conception: toward a theory of conceptual change, *Science Education* 66, pp 211-227.

Pouillet, C.S.M. 1828. *Eléments de physique expérimentale*, vol 1, 1st edition, p. 635.

Pouillet, C.S.M. 1847. *Eléments de physique expérimentale*, vol 1, 5th edition, p. 596.

Rainson, S. 1995. *Superposition des champs électriques et causalité: étude de raisonnements, élaboration et évaluation d'une intervention pédagogique en classe de Mathématiques Spéciales Technologiques* .Thesis, Université Paris 7.

Rainson, S., Tranströmer, G. and Viennot, L. 1994. Students' understanding of superposition of electric fields. *American Journal of Physics*, 62 (11), pp 1026-1032.

Rebmann, G. and Viennot, L. 1994. Teaching algebraic coding. *American Journal of Physics*, 723-727.

Roqueplo, P. 1974. *Le partage du savoir*. Seuil, Paris.

Rosser, W.G.V. 1970. Magnitudes of surface charge distributions associated with electric current flow. *American Journal of Physics*, n°38, pp 265-266.

Rozier, S. 1983. *L'implicite en physique: les étudiants et les fonctions de plusieurs variables*, Mémoire de D.E.A., Université Paris 7, L.D.P.E.S.

Rozier, S. 1988. *Le raisonnement linéaire causal en thermodynamique classique élémentaire*. Paris, Thesis, Université Paris 7.

Rozier, S. and Viennot, L. 1991. Students' reasoning in elementary thermodynamics. *International Journal of Science Education,* 13 (2), pp 159-170.

Rumelhart, R.D. and Norman, D.A. 1978. Accretion, tuning and restructuring: three modes of learning. In *Semantic Factors in Cognition*, J.W. Cotton and R. Klatzky, Lawrence Erlbaum Associates: Hillsdale, NJ.

Saison, A., Malleus, P., Huber, P. and Seyfried, B. 1979. *Physique 4ème*. Paris, Nathan.

Saltiel, E. 1978. *Concepts cinématiques et raisonnement naturels: étude de la compréhension des changements de référentiels galiléens par les étudiants en sciences*. Thesis (Thèse d'état),Université Paris 7.

Saltiel, E. and Malgrange, J.L. 1979. Les raisonnements naturels en cinématique élémentaire. *Bulletin de l'Union des Physiciens*, 616, pp 1325-1355.

Saltiel, E. and Viennot, L. 1983. *Questionnaires pour comprendre*, Université Paris 7 (diffusion L.D.P.E.S.).

Saltiel, E. and Viennot, L. 1985. What do we learn from similarities between historical ideas and the spontaneous reasoning of students? *The Many Faces of Teaching andl Learning Mechanics*. In Lijnse, P. ed. GIREP/SVO/UNESCO, pp 199-214.

Saltiel, E. 1989. Les exercices qualitatifs fonctionnels. *Actes du colloque sur Les Finalités des Enseignements Scientifiques*. Marseille, pp 113-121

Sanner, M. 1983. *Du concept au fantasme*. Paris, P.U.F.

Schiele, B. 1984. Note pour une analyse de la notion de coupure épistémologique. in Schiele, B., Belisle, C. and Garnier C. (eds) *Communication-Information: les représentations*. 6 (2-3). CIRADE, Montréal pp 43-98.

Séré, M.G. 1982. A propos de quelques expériences sur les gaz: étude des schèmes mécaniques mis en oeuvre par les enfants de 11 à 13 ans, *Revue Française de Pédagogie*, 60, pp 43-49.

Séré, M.G. 1985. *Analyse des conceptions de l'état gazeux qu'ont les enfants de 11 à 13 ans, en liaison avec la notion de pression, et propositions de stratégies pédagogiques pour en faciliter l'évolution.* Thesis (Doctorat d'état), Université Paris 6.

Shipstone, D. 1985. Electricity in simple circuits. In Driver, R., Guesne, E. and Tiberghien, A. (eds) *Children's Ideas in Science*. Open University Press, Milton Keynes, pp 33-51.

Thenard, L.J. 1813. *Traité élémentaire de chimie*, vol 1, 1st edition.

Tiberghien, A. 1984a. Revue critique sur les recherches visant à élucider le sens de la notion de lumière chez les élèves de 10 à 16 ans, *Recherche en Didactique de la Physique, Les actes du premier atelier international, La Londe les Maures, 1983*, CNRS, Paris, pp 125-136.

Tiberghien, A. 1984b. Revue critique sur les recherches visant à élucider le sens des notions de circuits électriques pour des élèves de 8 à 16 ans, in

Recherche en Didactique de la Physique, Les Actes du premier Atelier International: la Londe les Maures 1983, CNRS, Paris, pp 91-107.

Tiberghien, A. 1984c. Revue critique sur les recherches visant à élucider le sens des notions de température et chaleur pour des élèves de 10 à 20 ans. In *Recherche en Didactique de la Physique, Les Actes du premier Atelier International: la Londe les Maures 1983*, CNRS, Paris, pp 55-74.

Tiberghien, A. and Delacôte, G. 1976. Manipulations et représentations de circuits électriques simples par des enfants de 7 à 12 ans. *Revue française de pédagogie*, n° 34, pp. 32-44.

Tiberghien, A., Jossem, E.L. and Barojas, J. (Eds). 1988. *Connecting Research in Physics Education with Teacher Education*, HYPERLINK http://www.physics.ohio-state.edu/ ~jossem/ICPE/BOOKS.html.

Tonnelat, M.A. 1971. *Histoire du principe de relativité*. Flammarion, Paris.

Valentin, L. 1983. *L'univers mécanique*, Hermann, Paris.

Varney, R.N. and Fischer, L.H. 1980. Electromotive force: Volta's forgotten concept. *American Journal of Physics*, n° 48, pp 405-408.

Viennot, L. 1979. *Le raisonnement spontané en dynamique élémentaire*, Hermann, Paris.

Viennot, L. 1979. Spontaneous reasoning in elementary dynamics, *European Journal of Science Education*, 2, pp 206-221.

Viennot, L. 1982a. L'action est-elle bien égale (et opposée) à la réaction?, *Bulletin de l'Union des Physiciens.*, n° 640, pp 479-488.

Viennot, L. 1982b. L'implicite en physique: les étudiants et les constantes. *European Journal of Physics*, vol 3, pp174-180

Viennot, L. 1983. Pratique de l'algèbre élémentaire chez les étudiants en physique. *Bulletin de l'Union des Physiciens, n°* 622, pp 783-820.

Viennot, L. 1985. *Mécanique et énergie pour débutants*, Université Paris 7 (diffusion LDPES).

Viennot, L. 1987. Recherche en didactique autour de la transition Secondaire-Supérieur. *Bulletin de l'Union des Physiciens* 699, pp 1251-1268

Viennot, L. 1989a. Bilans de forces et lois des actions réciproques. *Bulletin de l'Union des Physiciens* n° 716, pp 951-970.

Viennot, L. 1989b. Obstacle épistémologique et raisonnement en physique: tendance au contournement des conflits chez les enseignants. in N.Berdnaz et C.Garnier (eds), *Construction des savoirs, obstacles et conflits.* CIRADE, Montréal, pp 117-129.

Viennot, L. 1992. Raisonnement à plusieurs variables: tendances de la pensée commune , *Aster*, n° 14, pp 127-142.

Viennot, L. 1993. Temps et causalité dans les raisonnements des étudiants, *Didaskalia* 1, pp 13-27.

Viennot, L. 1994. Recherche en didactique et nouveaux programmes d'enseignement: convergences. Exemple du programme de Physique de quatrième 1993 en France, *Didaskalia,* 3, pp 119-128.

Viennot, L. and Rainson, S. 1992. Students' reasoning about the superposition of electric fields, *International Journal of Science Education,* 14 (4), pp 475-487.

Viennot, L. and Rainson, S. 1999. Design and evaluation of a research-based teaching sequence: The superposition of electric fields, *International Journal of Science Education*, 21 (1), 1-16.

Volta, A. 1801a. Lettre du professeur Volta à J.C. de la Mètherie sur les phénomènes galvaniques. *Journal de Physique*, n° 53, pp 309-316.

Volta, A. 1801b. De l'électricité dite galvanique. *A.C.*, n° 40, pp 223-256.

Volta, A. 1802. Lettre de Volta sur l'identité du fluide électrique avec le prétendu fluide galvanique, à M. Bancks. *Journal de Chimie*, n° 2, pp 158-169.

Subject index

Name Index

1-MONTH